Management of Recreation and Nature Based Tourism in European Forests

Ulrike Pröbstl · Veronika Wirth · Birgit Elands ·
Simon Bell
Editors

Management of Recreation and Nature Based Tourism in European Forests

ESF provides the COST Office through a European Commission contract

COST is supported by the EU RTD Framework Programme

Springer

Editors
Dr. Ulrike Pröbstl
University of Natural Resources
and Applied Life Sciences
Institute of Landscape Development
Recreation and Conservation Planning
Peter-Jordan-Strasse 82
1190 Vienna
Austria
ulrike.proebstl@boku.ac.at

Veronika Wirth
University of Natural Resources
and Applied Life Sciences
Institute of Landscape Development
Recreation and Conservation Planning
Peter-Jordan-Strasse 82
1190 Vienna
Austria
veronika.wirth@boku.ac.at

Dr. Birgit Elands
Wageningen University &
Research Centre
Forest & Nature Conservation Policy Group
Droevendaalsesteeg 3
6708 PB Wageningen
The Netherlands
birgit.elands@wur.nl

Dr. Simon Bell
Edinburgh College of Art
OPENspace Research Centre
Lauriston Place
Edinburgh
EH3 9DF
UK
s.bell@eca.ac.uk

ISBN 978-3-642-03144-1 e-ISBN 978-3-642-03145-8
DOI 10.1007/978-3-642-03145-8
Springer Heidelberg Dordrecht London New York

Library of Congress Control Number: 2009939416

Cover illustration: Jari Kostet (Metsähallitus)

Cover design: deblik, Berlin

Printed on acid-free paper

Springer is part of Springer Science+Business Media (www.springer.com)

Cost

COST – the acronym for European Cooperation in Science and Technology – is the oldest and widest European intergovernmental network for cooperation in research. Established by the Ministerial Conference in November 1971, COST is presently used by the scientific communities of 35 European countries to cooperate in common research projects supported by national funds.

The funds provided by COST – less than 1% of the total value of the projects – support the COST cooperation networks (COST Actions) through which, with EUR 30 million per year, more than 30,000 European scientists are involved in research having a total value which exceeds EUR 2 billion per year. This is the financial worth of the European added value which COST achieves.

A "bottom up approach" (the initiative of launching a COST Action comes from the European scientists themselves), "à la carte participation" (only countries interested in the Action participate), "equality of access" (participation is open also to the scientific communities of countries not belonging to the European Union) and "flexible structure" (easy implementation and light management of the research initiatives) are the main characteristics of COST.

As precursor of advanced multidisciplinary research COST has a very important role for the realisation of the European Research Area (ERA) anticipating and complementing the activities of the Framework Programmes, constituting a "bridge" towards the scientific communities of emerging countries, increasing the mobility of researchers across Europe and fostering the establishment of "Networks of Excellence" in many key scientific domains such as: Biomedicine and Molecular Biosciences; Food and Agriculture; Forests, their Products and Services; Materials, Physical and Nanosciences; Chemistry and Molecular Sciences and Technologies; Earth System Science and Environmental Management; Information and Communication Technologies; Transport and Urban Development; Individuals, Societies, Cultures and Health. It covers basic and more applied research and also addresses issues of pre-normative nature or of societal importance.

Web: http://www.cost.esf.org

Acknowledgements

This book is the result of a four year collaboration of 40 researchers and practitioners in the field of forest recreation and nature based tourism. The basis for this publication are the shared experiences from participants from 26 European countries collaborating in the working group "Recreation Planning and Management" within COST Action E33 "Forest recreation and nature based tourism". COST Office and the European Science foundation supported our meetings twice a year in different European locations. The editors would like to thank the staff of COST Office at Avenue Louise in Brussels for their facilitation of the meetings and their administrative support. Special thanks go to the successive scientific officers who looked after the Action – Arne Been, Günther Siegel and Melae Langbein for their guidance and help. We would also like to thank the team of the publication office at COST for their support with the administrative aspects of the publication. Finally we would like to acknowledge Springer Verlag for their professional guidance during the compilation of the book. The editors of the book would like to express their serious gratitude to all working group members for their enthusiasm, inspiration and also their hard work that made this book possible. The entire four years of COST Action E33 were a gratifying and rewarding experience.

Ulrike Pröbstl, Veronika Wirth, Birgit Elands and Simon Bell
July 2009

Contents

Chapter 1
Introduction

Ulrike Pröbstl, Veronika Wirth, Birgit Elands, and Simon Bell

1.1 Background to the Study

Ulrike Pröbstl and Simon Bell

New directions and the need for reform - these two catch phrases dominate the political discussions around forests, forestry, and forest research in many European countries. In Germany there is concern about the reduction of university positions in forestry, as there is across Europe about the new, more economically orientated direction of forest management. For example, in 2004 the Bavarian electorate defeated a referendum about the restructuring of the forest administration, largely because of concern about changes to nature conservation and recreation management. Obviously, the new trends in forest management imposed by commercial interests stand in stark contrast to the overall expectations society has of effective multi-functional forest management (Deutscher Rat für Landespflege 2004). The administrative desires for reform on the one side and - often contradictory - societal demands on the other are the reason to undertake an investigation of the European model of multi-functional forest management.

The situation in Germany is reflected in many regions in Europe - the countries of the former Soviet Union and Eastern Bloc who recently joined the EU or who have aspirations to do so, have made great changes in the way that forests are owned and managed, in the way that forestry and timber production contributes to the national economy and the role and importance of nature conservation and recreation. European forests, forest culture and traditions of use for all sorts of timber and non-timber forest products are very diverse. Some countries are heavily forested and have low population density (for example Finland), while others have small amounts of forest and high population density (for example the Netherlands). This is reflected in the relative importance given to different forest functions, in particular recreation

U. Pröbstl (✉)
University of Natural Resources and Applied Life Sciences, Vienna, Institute of Landscape Development, Recreation and Conservation Planning, Peter-Jordan-Strasse 82, 1190 Vienna, Austria
e-mail: ulrike.proebstl@boku.ac.at

U. Pröbstl et al. (eds.), *Management of Recreation and Nature Based Tourism in European Forests*, DOI 10.1007/978-3-642-03145-8_1,

versus timber production. In some countries, such as the Baltic States, there has been a transformation of ownership from 100% state owned in Soviet times to something approaching 50% private ownership, as land has been restored to its former owners. This has a profound effect on recreational possibilities, especially near cities. There are also major differences in accessibility, ranging from "Everyman's right" in the Nordic countries to non-accessible private forest in others. Thus a European overview has to take these different perspectives into account.

Forest functions have changed over time. The various social functions are becoming increasingly recognised as being very important and include not only recreation but also health and well-being and wider quality of life. Also in this field new challenges arose such as

- changing recreational traditions
- demographic change
- new forms of tourism and recreation
- changing importance of wood production.

In some countries recreational activities such as seasonal collections of berries and mushrooms have always been an integral part of rural life. Visits by urban-based people to forests for walking and picnicking also have a long tradition. Tourists visiting other countries frequently visit forests and may camp or stay in cottages or other accommodation in the forest. However, the demands for forest recreation have been increasing in volume and have become more diverse. Furthermore in some areas there are severe pressures on many forest areas and they are in danger of being degraded ecologically and physically due to high recreational use.

The demographic structure of the European population is changing. In particular the ageing population presents special challenges: the "new" elderly are likely to have more time and money, but to be less fit and to have special requirements. Also, the requirements of poorer people may be different from those of wealthier citizens. They are more likely to need facilities that are located in places they are able to travel to easily. This is a part of the recent political trend to address social exclusion (Bell et al. 2007).

Tourism within and between European countries is changing as new areas become popular and new types of tourism activity are developed. The natural character of forests means that they are seen as providing important benefits to operators of nature tourism, a growing market at the present time. New challenges derive from the demographic change, new forms of land use like nature based tourism but also new conflicts.

While holiday destinations further afield are becoming popular to those who can afford them, the forests of Europe continue to offer attractive experiences, but only as long as their qualities are retained. Exploitation of forests for timber production may in some cases have negative effects on the qualities of the landscapes in which recreation takes place. Poorly developed and maintained facilities may devalue the very qualities that people have come to see and over-use may result in environmental degradation. Thus, forest recreation and tourism has to be sustainable at both the site level and over the forest as a whole.

With the regionally changing importance of wood production in many European regions, and major structural changes to the agricultural economy new challenges and tradeoffs are required. In many eastern European countries nature tourism is looked upon as a means for rural development. Nature based tourism is no longer merely the provision of basic tourism services but increasingly giving the clients possibilities for different kinds of experience. The quality of the environment is one of the elements defining the quality of a nature tourism service.

Due to its increased importance and complexity, as well as the potential conflicts with other functions, the social function of forests has to be dealt with much more explicitly than in the past. This increased attention and awareness is required at different levels, from policy-making, spatial planning, forest and site design to the management of a specific area.

Recreational aspects of forests have been mostly studied at a national level. The literature therefore tends to be fragmented and dispersed. This publication provides a first Europe-wide overview on research, education and management in the field of forest recreation and nature tourism. It contains a scientifically-based description of the diverse "forest cultures" of Europe, where the relative amounts and types of forest, the traditional uses made of them and the place of forests in the culture of the countries differ, and presents their fascinating diversity in a truly comparative manner. These regional, as well as national differences need to be celebrated and strengthened, because they allow visitors from other countries to experience something different.

This study was carried out within the framework of the COST Action[1] E33 "Forests for Recreation and Nature Tourism (FORREC)"[2] by the working group on recreation planning and management. In FORREC, researchers and policy-makers of 27 different European countries co-operated to understand and improve the benefits of all kind of issues related to the role of forests for society in general and for recreation and tourism in particular. While most of Europe was represented in the action, there were several notable absences that left holes in the analysis - Spain, Slovenia, Romania, Bulgaria, Macedonia and Albania did not join the action at all and Estonia joined too late for the data to be collected.

The aim of the book is to give an overview of the state of the art, traditions, differences and similarities in the use and management of recreation and nature tourism in European forests. Furthermore the book shows new tasks, goals and strategies. The final discussion provides a profound insight into future trends regarding forest recreation and nature based tourism. It is structured around a series of different European regions with different patterns of forests, traditions of recreation and different problems and concerns. Examples of good practice from different situations are also given and the literature provided for reference purposes.

---

[1] http://cost.cordis.lu/src/home.cfm

[2] http://www.openspace.eca.ac.uk/costE33/welcome.htm

1.2 Structure of the Book

Ulrike Pröbstl, Veronika Wirth and Birgit Elands

1.2.1 Research Design

At the beginning of the Cost Action E33 in 2004, initial discussions with the group of European experts showed that there is no one single approach or form of good practice for recreation management in forests; instead a suite of different problems and experiences exists, to which a variety of management solutions have been adapted in different regions. This experience at the beginning of this European exchange of knowledge was the starting point for an innovative research initiative to explore similarities and differences across Europe. A structured European-wide survey of forest recreation and nature tourism was undertaken. The aim of the survey was to provide a general overview of the European situation of forest recreation and nature tourism, reviewing the current state of knowledge and methods regarding recreation planning and management. The aim was to give a broad perspective of conflicts and management concepts and to identify differences and similarities.

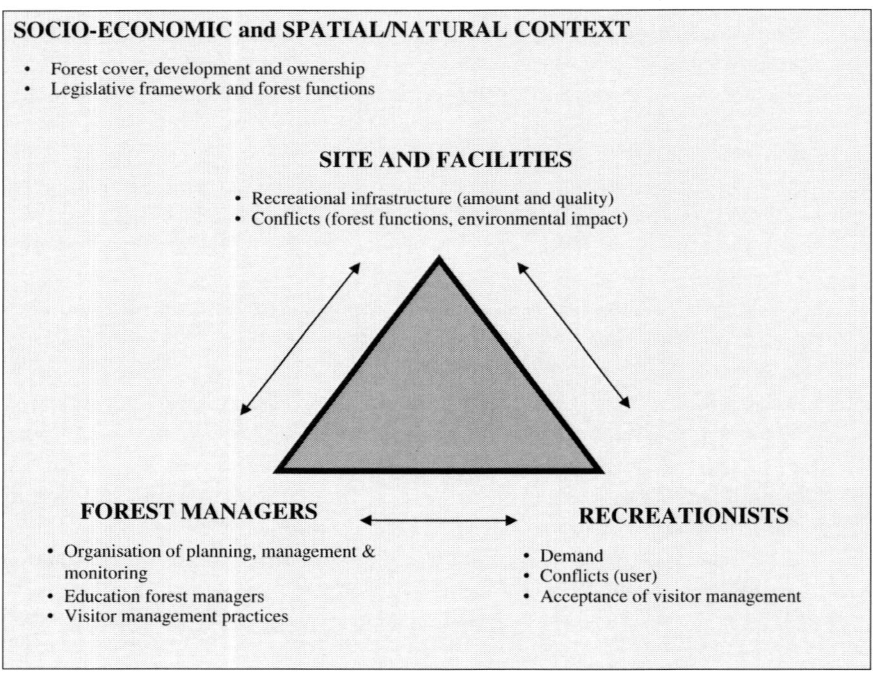

Fig. 1.1 Conceptual framework for the Europe-wide survey

For the expert based study a questionnaire was developed with the following main subjects (see also Fig. 1.1):

- *socio-economic and spatial context of each country*: forest cover, development and ownership and legislative framework and forest functions;
- *site and facilities*: recreational infrastructure and conflicts with respect to different forest functions and environmental impact;
- *forest management*: organisation of planning, management and monitoring, education of forest managers, and actual visitor management practices;
- *recreationists*: demand, conflicts between recreationists and types of activities, and acceptance of visitor management measures implemented by forest managers.

Any expert based study on a European level faces the following problems: There is a wide variety of languages (in this case involving 27 countries speaking 22 different languages) and the selection of suitable experts in theory and practice is difficult.

In this particular case the experts from the working group "Recreation Planning and Management" within Cost Action E33 formed the basis for the collection of the data and they chose further national experts in each of the 27 countries (see Fig. 1.2).

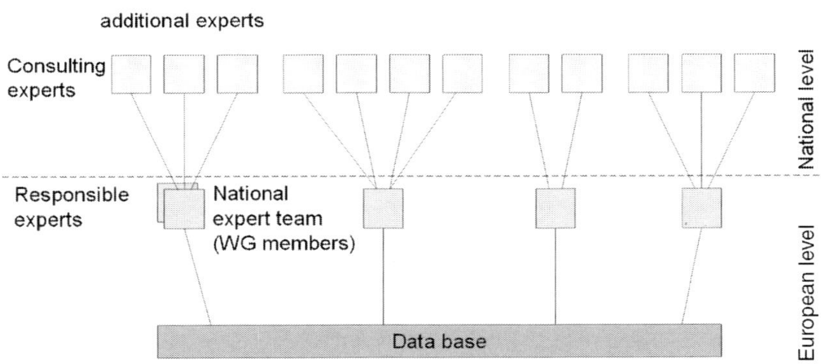

Fig. 1.2 Scheme of the expert based approach used in this study. (The responsible experts were members of the working group of the cost action E33)

From each country one or two national experts were involved. There were limitations placed on what could be achieved due to the funding of the Cost Action (travel to meetings is supported by Cost funding but staff time has to be funded by the participants' home institutions which in several cases limited how much individuals could contribute) but the country experts provide a diverse cultural and historic background as well as range of experiences and quality of input regarding forests recreation and nature tourism. The experts can be distinguished by the following characteristics:

- *Disciplinary background*: forestry, biology, landscape architecture, sociology, and land use planning among the participants;
- *Working environment*: university, research institute, forestry service, national/regional government, NGO, and private consultancy.

The questionnaire was addressed to the national representative member, who was invited to be responsible for the questionnaires in his or her country, consulting additional experts based mainly on a personal invitation. This led to a very high response rate.

The method has some clear advantages. The expert based questionnaire makes use of various networks of experts. Another advantage of this method was that there is no language barrier. At the international level, within the working group the language was English, whereas at the national level the national languages could be used. The method has also some weak points: there were some minor variations in the estimations and ratings within some countries and between countries in the same region (see below).

The analysis shows that the socio-economic and spatial/natural conditions are framing and dominating the national results. This leads to regionally connected clusters with similar conditions. These effects can be illustrated by the following two examples:

- In Cyprus tourism is one of the main economic factors. Conflicts with tourist facilities in forests for tourists are as important as the natural and man made danger of forest fires. Similar demands and requirements for forest management exist in Greece, Portugal or Croatia. The concepts to solve these problems, however, may be different.
- The forest use in the Netherlands is strongly influenced by dense population and a small amount of forests. This together with the important conservation function of the remaining forests lead to difficult management tasks which also can be found in the United Kingdom or Denmark.

This was the reason for organizing the interpretation of the results by regional groups (see the following Chapters 2–6). The national findings were discussed in the working group as a whole and split up into regional groups with the possibility to make revisions and adjustments as necessary.

The results included the answers from practitioners, researchers and foresters working within the respective regional or national administration. Only in one country with different regional conditions were there difficulties to harmonize the results to create one single national response. The representatives reported that in most cases the main findings of the different experts corresponded well with each other. The conclusions drawn from the survey are of more general kind and focus on the European comparison. They do not go into detailed description of countries' internal differences.

The interpretation of the results was carried out together with the representatives and additional local or regional partners to include and explain the background and possible regional differences. Data was analysed both quantitatively and qualitatively by the editors.

All the involved experts were finally invited to illustrate their management experience and scientific findings by good or best practice in their regions (see Chapter 8).

1.2.2 Regional Approach

The similar socio-economic and natural conditions, such as the overall amount of forests present, the population density, legal foundations, and especially the conditions of access to forests have led to the identification of regional groups of countries regarding forest recreation and management in Europe (see Fig. 1.3).

The *Atlantic region* covers the countries of Belgium, Denmark, Iceland, Ireland, the Netherlands and the United Kingdom. Their main characteristics are the lowest forest cover in Europe and a high population density and thus a very high frequency

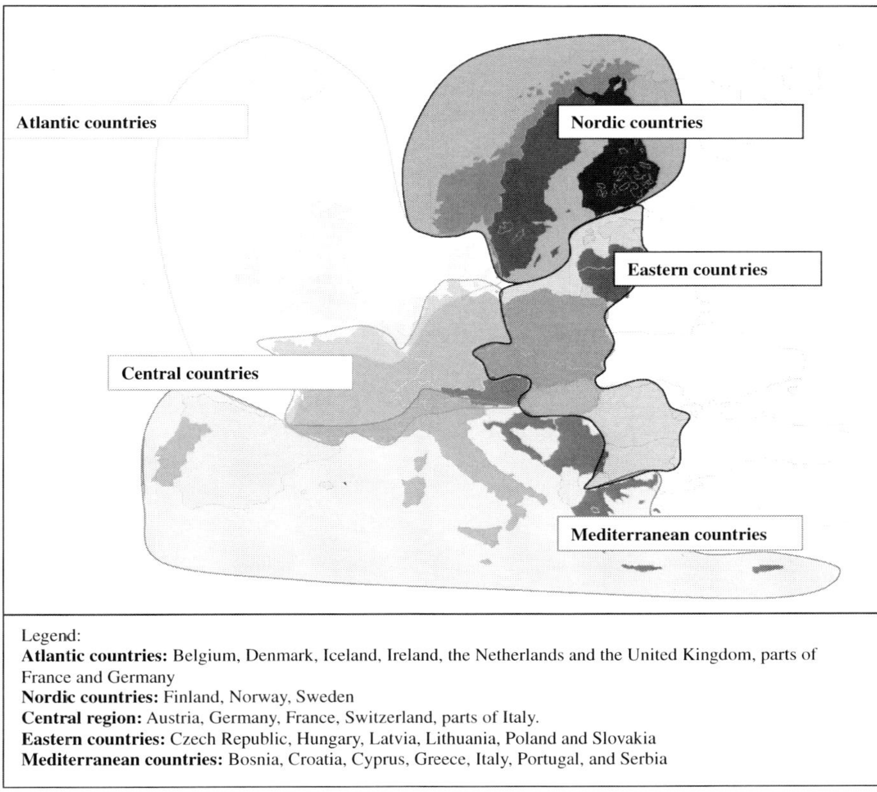

Legend:
Atlantic countries: Belgium, Denmark, Iceland, Ireland, the Netherlands and the United Kingdom, parts of France and Germany
Nordic countries: Finland, Norway, Sweden
Central region: Austria, Germany, France, Switzerland, parts of Italy.
Eastern countries: Czech Republic, Hungary, Latvia, Lithuania, Poland and Slovakia
Mediterranean countries: Bosnia, Croatia, Cyprus, Greece, Italy, Portugal, and Serbia

Fig. 1.3 Regional clusters of countries within Europe

of recreational use in the forests compared with the rest of Europe. Also, the Atlantic coastal part of France and Spain and the highly populated western parts of Germany show some characteristics of this region although they were not specifically included in the analysis. The high population density and relatively low proportion of forest (<11%) leads to an emphasis on recreation in research and practice.

In the *Nordic region* Finland, Norway and Sweden are included. The main characteristics are the highest forest cover in Europe, a low population density and a free access to the forest for recreational purposes. There is the traditional right of "Everyman's right", which include berry picking, mushroom collecting, and free access for all recreation and sports activities, which are considered of the utmost importance by the entire population (Sievänen 2001a).

Countries within the *Central region* are Austria, France, Germany and Switzerland. The main characteristics are: a third of the country is covered by forests with a tendency for some abandoned areas to become reforested; mostly the mountainous regions (low mountain ranges and the Alps) are covered by forests. All countries provide free access to the forest. However, there may be minor restrictions regarding type of use (for example in Austria) and forest ownership (in France). The region has a rather high population density. Also the alpine part of Italy shows some characteristics of this region.

The *Eastern region* covers the countries Czech Republic, Hungary, Estonia, Latvia, Lithuania, Poland and Slovakia. Not all Eastern European countries have been involved in the COST Action E33 and are therefore not included in this study (see above). This region faces a different set of challenges. The main characteristics are that the region has been influenced by the former Communist regimes of the Eastern Bloc and Soviet Union. Over the last couple of decades there have been large privatisation campaigns which also strongly affected forest ownership. The restructuring of agriculture and the widespread conversion of abandoned farmland to scrub and eventually forest leads to significant changes in the rural landscape. The reforested areas will also pose new challenges to forest management in the future.

The Mediterranean region includes: Bosnia, Croatia, Cyprus, Greece, Italy, Portugal, and Serbia. One third of this region is covered by forest but here it is a rather different type of forest. There is a strong influence from the Mediterranean climate, making forest fires in the hot summers one of the main conflicts regarding recreation planning and management.

The regions are described in detail in the following chapters. With the regional approach adopted by this study it is possible to describe the conditions in different countries and to describe a landscape that belongs to several different countries (such as the Alps or the Mediterranean coast). It is also possible to make more general statements on the European situation of forest recreation and nature based tourism. Furthermore, comparisons between different solutions within a region may allow the adoption of solutions for joint problems or conflicts. Also, some - however limited - observations can be made about countries that have not been involved in the Cost Action.

The results of the survey will be described in Chapters 2–6 at regional and in Chapter 7 at European level. The following topics will be discussed within each regional chapter:

- Introduction;
- Forest cover, development and ownership;
- Legal background, public access and forest functions;
- Forest manager's education and tasks regarding recreation in forests;
- Conflicts;
- Planning, management and monitoring;
- Visitor management;
- Future trends and challenges.

Chapter 2
Atlantic Region

Birgit Elands, Simon Bell, Jan Blok, Vincent Colson, Sherry Curl, Berit C. Kaae, Gudrun Van Langenhove, Art McCormack, William Murphy, Jon Geir Petursson, Søren Præstholm, Pieter Roovers, and Roger Worthington

2.1 Introduction

Chapter 2 explores recreation and tourism practices in forest areas in the Atlantic region, which refers to the geographical area close to the North Sea and the Atlantic Ocean. The Atlantic countries described in this section are Belgium (Flanders and

Fig. 2.1 Countries within the Atlantic region

B. Elands (✉)
Wageningen University & Research Centre, Forest & Nature Conservation Policy Group, Droevendaalsesteeg 3, 6708 PB Wageningen, The Netherlands
e-mail: birgit.elands@wur.nl

U. Pröbstl et al. (eds.), *Management of Recreation and Nature Based Tourism in European Forests*, DOI 10.1007/978-3-642-03145-8_2,

Wallonia), Denmark, Iceland, Ireland, the Netherlands, and the United Kingdom (England, Scotland, Wales and Northern Ireland, although there is no specific data for Northern Ireland in the survey) (Fig. 2.1). In this chapter, instead of always focussing on the level of the country, we sometimes refer to a region within a country. This region might be largely political-administrative independent or geomorphologically distinct. Good examples are the regions of Flanders and Wallonia in Belgium and Scotland, England and Wales in the United Kingdom (UK), where there are separate administrations and forestry policies, organisations and support systems. We have chosen to describe them predominantly at the country level. Parts of north-western France, western Germany and south eastern Sweden could also be said to belong to the Atlantic region, but are assigned to the Central European and Scandinavian region respectively because there are no separate data available from the survey.

What makes the Atlantic region distinct from the rest of Europe? There are two main features common in this region, although these do not apply to all countries equally: (i) forest cover and history and (ii) degree of urbanisation and population density. The first distinctive feature is the fact that each country has a much lower percentage of land under forest than is the average for most other countries in Europe. These percentages range from an estimated 1.7% woodland cover in Iceland to 22.2% in Belgium. The forests in most cases are the result of afforestation programmes undertaken throughout the twentieth century. The reason that afforestation programmes are underway is because the natural forests that were once a major feature of all countries (even Iceland, which many people think of as a barren, windswept land) were cleared away over the last few thousand years or last few centuries. Thus many forests have a short history. A second distinctive feature of the Atlantic countries is its high population density and strong urbanisation. The Netherlands is most extreme with about 480 persons per km^2 and almost all inhabitants from Iceland live in urban areas. North-west France also has a low percentage of forest as does the Skåne region of Sweden, while the western-most part of Germany has a very high density of urbanisation and population, which is why these areas could also be considered part of the region.

This situation of high population density and low amounts of forest implies that forests are highly valued for recreation because, if they are open to public access (often the case with public or state forest but not necessarily in privately owned forests, of which there are many), they offer a chance to escape from the pressures of the city. Many have been designated as forest parks or something similar and have been equipped with visitor centres, car parks, picnic sites, trails and other facilities. Some of the state and public forests now also host privately operated tourism ventures, or provide venues for specific activities such as mountain biking, where local enterprises ranging from bed and breakfasts to bike rental shops and local restaurants provide income and employment in remote rural areas. Nature tourism is perhaps less well developed in the Atlantic countries and may be seen as less important than in some other European regions, but it has a role to play and presents its own challenges and conflicts for planning and management.

2.2 Forest Cover, Development and Ownership

2.2.1 Forest Cover and History

The amount of forest cover and the distribution of forest areas in relation to population clusters are important for the recreational opportunities in all of the countries. In the Atlantic countries, the forest cover (Fig. 2.2) varies from between under 2% in Iceland, to around 8% in Ireland and about 11% in Denmark, the Netherlands and the UK. Belgium has the highest forest cover (22%) with big differences between Flanders (11%) and Wallonia (32%).

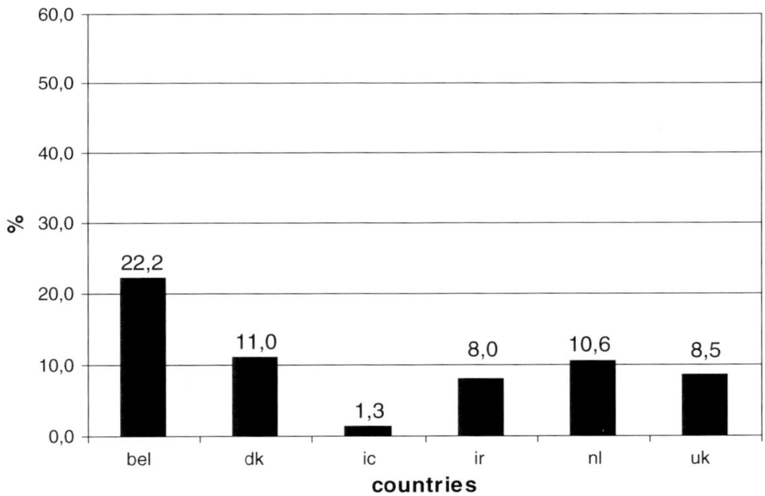

Fig. 2.2 Forest cover (%) in the Atlantic countries

The forests in most cases are of plantation origin and are the result of afforestation programmes that were undertaken mainly in the first half of the twentieth century but which are also continuing in most countries. The natural forests that were once a major feature of all countries disappeared as the result of a combination of factors – to make way for agriculture, to provide fuel or timber, or as the result of uncontrolled grazing animals eating the natural regeneration so as the old trees died, they were not replaced. In many cases some remnant forests remained, saved as a result of their use as royal hunting forests or because of their inaccessibility. Thus the New Forest in the south of England, Jaegersborg Dyrehaven to the north of Copenhagen in Denmark, the Mastbos near Breda in the Netherlands, and the Zoniënwoud/Forêt de Soignes outside Brussels in Belgium are relics of managed native forest full of many cultural and natural heritage, landscape and biodiversity values as well as being important recreational resources. Wallonia still has a lot of broad-leaved forests which have persisted for several centuries.

Initially, in most places, afforestation programmes were started to increase the available reserves of timber. In the UK, Ireland and Iceland conifer species from North America and Northern Europe were chosen as these grow faster and can also withstand the windy, wet and cool climates typical of these countries. In Denmark and Wallonia spruce was used together with other species, while the Netherlands and Flanders were afforested mainly with Scots pine, Japanese larch and Douglas fir. A feature of much of the afforestation was that the places chosen were frequently considered too poor for agriculture, so that in the UK and Ireland, for example, the poor quality remoter hilly and mountainous areas were chosen, where the soil was too infertile and wet to support farming. In Denmark heathland areas and areas too poor for agriculture were similarly planted, as was the case in the Netherlands and in Belgium. In Iceland a number of more sheltered valleys were initially used for experimental plantations, to test the feasibility of afforestation. In mostof these countries afforestation programmes remain active, although the objectives may have changed over time (see Section 3).

2.2.2 Forest and Demographics

Important for recreation and tourism is the relation of forest cover to population density. Figure 2.3 shows that the Atlantic region is densely populated. The single exception is Iceland with a low population density. Iceland, however, is an extremely urbanised country, in the sense that the majority (95%) of the population

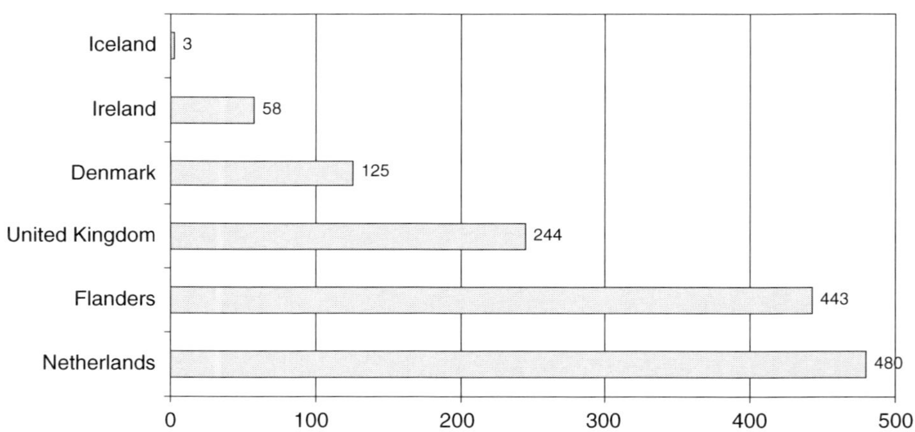

Fig. 2.3 Population density in Atlantic countries (number of persons/km^2)[1]

[01] Source: Eurostat (http://epp.eurostat.ec.europa.eu/portal/page/portal/eurostat/home)and Federale Overheidsdienst Economie, KMO, Middenstand en Energie (http://statbel.fgov.be/figures/dsp2003_nl.asp)

Table 2.1 Area of forest and other wooded land (FOWL) per capita

Country	FOWL/capita (ha/capita)
Netherlands	0.022
United Kingdom	0.038
Belgium	0.066
Denmark	0.102
Ireland	0.161
Iceland	0.471

live in urban areas. The UK, the Netherlands, Belgium and Denmark are the most urbanised, while regionally, Ireland has several major urban centres and is becoming increasingly more urban (over 70% of Irish people now live in “urban areas”). The UK also has some marked regional differences, with Scotland and Wales being less populated and parts of Scotland having a very low population that is on a par with Iceland in terms of density. However, this is compensated for by the high density in the south of England or South Wales. In Belgium, Wallonia (202 persons/km^2) is much less urbanised than Flanders (443 persons/km^2). Likewise, there are some small differences in the Netherlands, where the northern and eastern areas are less densely populated.

Table 2.1 presents the available area of forest and other wooded land per inhabitant for each of the Atlantic countries. Although Iceland has a very limited forest cover, the amount of forest and other wooded land per inhabitant is the largest. In Belgium, large within-country differences in both population density and forest cover imply that Flanders (0.024 ha/capita) resembles the Netherlands and Wallonia (0.16 ha/capita) resembles Ireland.

2.2.3 Forest Distribution

Regarding the distribution of forests throughout the country, three out of the six countries suggest that – although not distributed evenly throughout the country – forests can easily be reached by a majority of the population (two-thirds). Most forests are located close to urban areas. Despite the low forest cover in Iceland, the majority of its inhabitants live within acceptable travelling distance from the forests. The pattern of forest distribution, as a result of the afforestation programmes in the Netherlands and the UK, resulted in the forests being relatively far from the main population centres. This can be historically explained, as the forests were established primarily for timber on the poorest agricultural land far from urban centres (see above). Consequently, the situation in both countries is more pronounced, because the unevenly distributed forests can be reached only by a minority of the population (one-third). Only in Ireland are the forests equally distributed within the country resulting in a relatively even level of accessibility across geographical areas. The dispersed nature of forest holdings means that forests are relatively close to urban areas, including the capital, Dublin.

2.2.4 Forest Ownership

Forest ownership influences public accessibility and recreational opportunities of forests. The ownership patterns differ greatly across the Atlantic countries (Fig. 2.4). If a distinction is made between privately owned and publicly owned forests (community forests, state forests, etc.), it becomes obvious that forests in Ireland and Iceland are mostly publicly owned (although the extensive farm forestry programmes in both countries will significantly increase the area of forests privately owned in the future). In Belgium, and Denmark, most forests are privately owned. The Netherlands holds an intermediate position. Community owned woods (public woods) are significant features in Belgium, Iceland, and the Netherlands. In England, many urban forests are owned by public bodies, utilities and local authorities. The most recent afforestation projects are located on derelict brown-field sites (former industrial land).

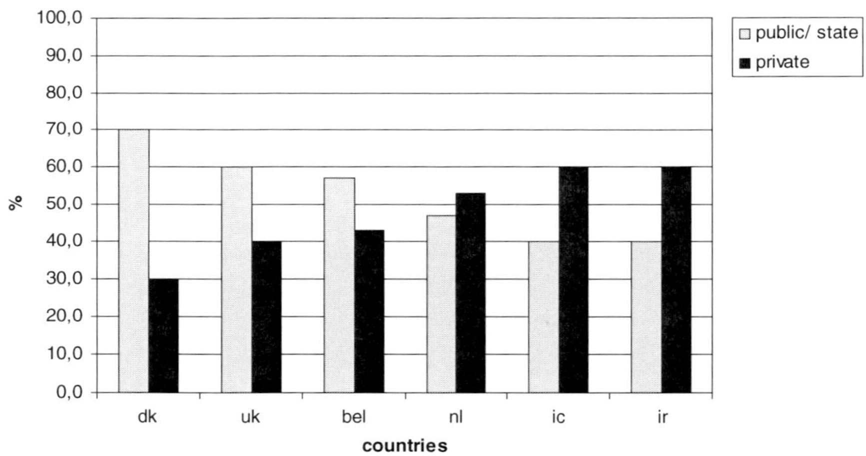

Fig. 2.4 Forest ownership in the Atlantic countries (%)

In all countries, small-scale forest-ownership prevails. For example, in Denmark 90% of the holdings are less than 20 ha and in Flanders 75% of the holdings are less than 2 ha. Occasional larger areas of forest also occur in parts of this region – for example some forests in the north of Scotland – but generally the forests are relatively small patches surrounded by farmland or, increasingly, near cities or other built up areas.

There are no clear regional differences regarding forest ownership, except in the UK, where there is a greater density of private woodlands in South East England, and in Iceland. In half of the countries, the size of the private forest holdings does not vary widely across the country. In Iceland and the Netherlands however, there are significant regional differences in the size of private forest holdings.

2.3 Legal Background, Public Access and Forest Functions

2.3.1 Legal Background

National legislation is a key issue in providing public access and recreational opportunities of forests to the public. Legally, the Atlantic countries have different backgrounds in terms of the accessibility of forests. There is a long tradition of open access in the countryside in Iceland but also in the UK. Until recently, while state forests were accessible it was only possible to use public rights of way to gain access to most private land. This is part of the cultural heritage of these countries. Recently a "right of public access to land" was written into national legislation:

- In England and Wales there is the CRoW (Countryside and Rights of Way) Act (2000). This Act introduces a statutory right of access on foot, for open air recreation, to land mapped as public access land. Scotland has also introduced a new statutory right of responsible access to land on foot for open air recreation (Land Reform (Scotland) Act 2003).
- In Iceland accessibility to nature areas is regulated in legislation (1999 nr. 44 22. mars) and allows open access for walking and hiking (non-motorized activities) in natural areas. However Iceland has limiting national legislation regarding planted forests (see below).

In the other countries, access is associated with ownership. In general, this implies that only publicly owned forests are freely accessible and privately owned forests are not accessible to the public. However, in those countries with high urban pressure and low forest cover, i.e. Belgium, Denmark, and the Netherlands, there are initiatives towards either introducing a statutory right of access on foot or encouraging private owners, through financial inducements, to make their property accessible to the public. In these countries, national legislation enables forest owners to influence public access for recreational use:

- In Denmark, in publicly owned forests, public access is permitted by foot to all parts of the forest 24 h a day. Since 1969 public access to roads and pathways in private forests has been permitted between the hours of 7am to sunset. Access by bicycle is permitted, but only on pathways.
- In Flanders, public access to all forests (private and public/state-owned) by foot on roads and pathways was incorporated in the Forest Act of 1990. Nevertheless, private forest owners can easily reverse the access by posting a sign "access forbidden, private property" at the main entrances. In Wallonia, access is allowed on every public road and trail (even if it is a public road through a private forest). Restrictions exist for specific user categories on some types of roads. For private roads and trails, it is up the owner to allow or deny public access.

- In the Netherlands, most state and publicly owned forests and nature areas are open for so-called "quiet recreational activities" on roads and pathways between sunrise and sunset (sometimes for 24 h a day). There is a strong tendency to encourage private forest and nature area owners to provide access by increasing or expanding economic incentives.

In Ireland and Iceland, limited national legislation exists that encourages forest owners to allow public access for recreational use. The state forest sector (managed by Coillte Teoranta) in Ireland allows and encourages permissive access on foot to its estates. The Irish Occupiers Liability Act (1995) allows for some recreational use of private forests and farmland and removes obstacles regarding liability associated with recreational use. Moreover, the Irish Forest Service is developing recreation guidelines to encourage recreational use of private woodlands. In Iceland, current legislation allows public access on foot to privately owned land in general, as long as there is no ongoing cultivation that could be damaged, which includes young forest plantations.

However, restrictions (prevention of access, limited access, entrance fees, access limits to some parts by means of barriers) may be implemented. In Denmark restrictions can be implemented in private forests smaller than 5 ha during forest operations and in the hunting season. Restrictions implemented in public and state forests are confined to the hunting season. In Belgium and the Netherlands, private owners can simply close their property (except public roads and trails going through their land) to the general public. Public and private owners of both countries can close (parts of) their estates for specific recreational activities (e.g. noisy activities that disturb the silence) or temporarily for reasons such as nature conservation (e.g. bird breeding season), hunting activities, timber harvesting, and times of high risk of fire. Specific owners, such as the Ministry of Defence or water collection companies, often have strong rules limiting access because of protection considerations.

In all countries, except Iceland, the size of the forest holding influences the type or degree of public access for recreational use, especially in private forests.

Another important issue is that statutory rights, such as those in Belgium, Denmark, the Netherlands, and the UK, should place some responsibilities on the visitors, such as access codes (e.g. Scottish Outdoor Access Code). Quite often, access rights are based on the public respecting general rules of proper behaviour in the forest. Signs with these rules are often posted at forest entrances (both public and private forests – where the rules may differ slightly) as a public reminder. In the Netherlands, owners generally work with "codes of conduct" and there are also good examples of codes of best practice for specific user groups (e.g. best practice code for mountain bikers). In Ireland the state forestry company (Coillte Teoranta), the Forestry Service and others such as the National Park Service and Heritage Council are developing a "Leave No Trace" programme in partnership with outdoor recreation users and youth organisations, on an all Ireland basis (i.e. including Northern Ireland) (Anon., 2005).

2.3.2 Impact of Legislation on Different Types of Forest Related Outdoor Activities

As forests must provide a range of different functions related to recreational activities, some countries have established regulations to avoid conflicts between forest users, e.g. the regulation of motorised activities to protect users visiting the forest to enjoy the peace and quiet, view nature, etc. Table 2.2 presents an overview of legislation in relation to different types of recreational activities in the Atlantic countries. Ireland is an exception, as there is no legislation regarding these activities.

The state forest company in Ireland is currently enacting by-laws to regulate activities such as mountain biking and motorised sport, but is encouraging this through self regulation using the Leave No Trace programme. Restricting behaviour by law does not guarantee compliance; therefore legislation should work together with enforcement and also in more positive ways, such as codes of best conduct for different user groups.

Activities such as camping or barbecuing are usually forbidden, unless the user has the permission of the forest owner. Open fires for barbecues can endanger the other functions of the forest by causing fires producing smoke and charred areas. Camping can disturb both flora and fauna. Horseback riding, mountain biking and orienteering are in most cases allowed, although some countries restrict these activities due to the disturbance and physical damage to the forest floor. In most countries, the extraction of forest products such as picking berries and mushrooms, which is considered an integral part of forest visits, is allowed for the visitors' own consumption. Only Flanders and the Netherlands do not allow the removal of anything from the forest.

Although legislation is country dependent, there is a negative correlation between the degree of legislation concerning outdoor activities and the amount of forest and wooded land (FOWL)/capita. Iceland and Ireland with the highest FOWL/capita have the least restrictions, whereas Flanders and the Netherlands with a low FOWL/capita have a greater amount of overall legislation. Denmark, Wallonia and the UK hold an intermediate position. This implies that more legislation is generally considered to be necessary as user intensity increases.

2.3.3 Forest Functions

Forests provide a number of different functions for society (timber production, protection of biodiversity, recreation etc.) and the non-timber related functions have generally increased in importance during the last 20 years. More recently there have been changes to the forestry policy and management in most, if not all, of these countries, moving away from timber production towards multiple objectives. Two of the most important functions are biodiversity and recreation.

Table 2.2 Impact of legislation on a variety of forest related outdoor activities

	BE	DK	IC	IR	NL	UK
Motorised recreation	Fl: prohibited by law W: only allowed on cobbled or tarmac roads or with temporary authorization (and conditions)	Prohibited by law	Prohibited by law	Under licence, bylaws control activity	Not allowed in state/public forests, sometimes in private forests	Allowed on statutory byways, otherwise prohibited by law of trespass on private land (pr/pu) and bylaws on state land (st)
Camping	Fl: prohibited by law and exceptional with permission of the forest service and the owner W: only allowed on certain areas	Prohibited by law	Allowed, but only with permission of the owner	With permission, bylaws control activity	Not allowed in state/public/private forests, unless regulated by local authorities	See motorised recreation
Barbecue	B: only allowed in designated areas	Prohibited by law, but sights are provided in st/pu forests	Without legislation	Permitted by consent – bylaws control activity	Depends on owner and on condition that it is regulated by local authorities	Allowed. pr/pu/st – permitted by consent (and use!)
Horse riding	Fl: only allowed on trails marked for horse riding W: allowed excepted on small trails (paths) which are not marked for riders	Allowed. Only st/pu forests	Conditional allowed	with permission of owner, bylaws control activity	+ state/public forests, +/– private forests, depends on owner	Allowed on statutory rights of way; permissive rights on designated tracks; otherwise restricted by trespass/bylaws as above
Mountain biking	Fl: only allowed on trails marked for bikers W: allowed excepted on small trails (paths) which are not marked for bikers	Only on forest roads and pathways	Conditional	In designated areas only, bylaws control activity	State/public/private forests, depends on owner	See horse riding

Table 2.2 (continued)

	BE	DK	IC	IR	NL	UK
Orienteering	B: prohibited unless by permit in designated areas	Allowed only in st/pu forests, > 30 persons or commercial activities demand permission	Allowed	In designated areas and by permit in undesignated areas	State/public/private forests, depends on owner	Allowed. Permit required
Mushroom picking	Fl: prohibited by law W: allowed for own consumption in Region forests and in the others only with the agreement of the owner	Allowed. For private use only	Allowed. For own consumption	No permit required	Prohibited by law (national Flora and Fauna Law)	Allowed. Private land restricted by law of trespass
Berries picking	Fl: not allowed W: allowed for own consumption in region forests and in the others only with the agreement of the owner	For private use only	For own consumption	No permit required	National Flora and Fauna Law	Public/state might be restricted by bylaw (no commercial picking) or by request

The social functions (recreation is one of these) are very important for all of the Atlantic countries – in particular countries with a high population density and intensively used landscapes. For the UK, forests play an increasing role in the provision of recreation, health and urban regeneration, especially in southern England where the forest's economical role is diminishing. In many parts of the UK, as well as in Belgium and Denmark, the social functions are considered to be as important as the other functions; consequently, they introduced a multifunctional management.[2] In Wallonia, forests are considered to be multifunctional but timber production is the most important function in many forests and the social function is only relevant in the peri-urban ones or forests located in tourism areas. In the Netherlands, recreation is the second most valued function (after biodiversity). Despite this overall Dutch difference, the social functions are of major importance in the forests managed by the National Forest Service (Staatsbosbeheer), who claim to use a multifunctional management plan. For Iceland, the social functions are also of high importance in both urban and rural forests. Although the social function is important for urban consumers, often it is less so for the private owners, who are experiencing great pressure to provide access and facilities.

The management of the existing forests has changed, especially in state or public forests: including in some cases major redesign and restructuring of the forests to improve them for landscape, biodiversity and recreation. Most examples of these types of programmes are probably found in the UK, and especially England, and the Netherlands where there has been a shift in the approach to afforestation. More forests or smaller areas of woodland are being planted in and around cities, closer to where people live, providing them with access to recreation on a daily basis, instead of having to drive some distance to get there. This is a significant feature of the forestry policy of the UK, the Netherlands, Denmark and Flanders. This approach pioneered mainly in the Netherlands, with the establishment of the Amsterdamse Bos in the 1930s. This is often hailed as an exemplar and provided inspiration to many countries. The Vestskoven in Copenhagen established as part of the "Fingerplan" from the 1960s is another example, and the Community Forest programme in England has been operating for around 17 years. In Iceland, a significant step was taken in this direction in 1950 when a substantial area (2,000 ha) was allocated for major afforestation programme to develop a forest recreational area in Heiðmörk for the population in and around the Reykjavík metropolitan area. Ireland developed a model for community forests – the Neighbourwood Scheme – to promote the development of "urban forests", but so far with limited success.

2.4 Forest Managers' Education and Tasks Regarding Recreation in Forests

The education of foresters has a significant impact on how recreational issues are managed in forests. In most countries, except Belgium and Ireland, knowledge about planning and management of forest-based recreation is considered to be part of the

[2]Under the England Forest Strategy woodland managers are required to balance social and environmental benefits with economically viable woodland management.

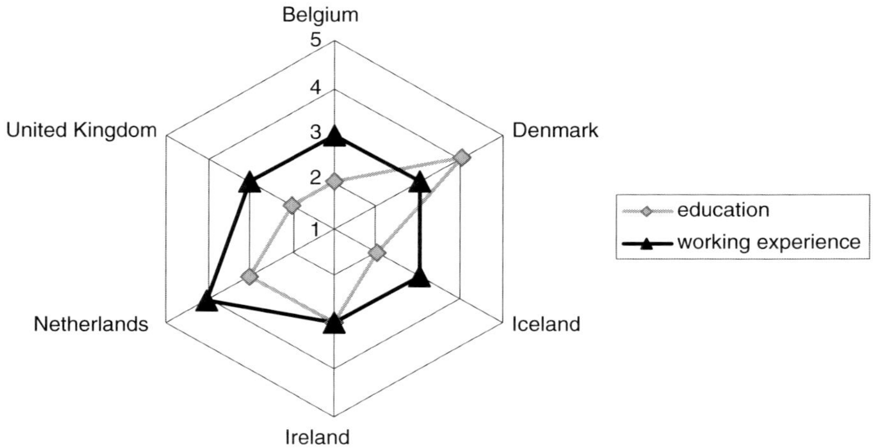

Fig. 2.5 Evaluation of knowledge about planning and management of forest based recreation obtained by education and working experience (1=unsatisfactory, 5=excellent)

basic knowledge and skills of foresters. This is achieved either through education or by working experience. These opinions are shown in Fig. 2.5.

This shows how experts estimate the importance of both education and working experience for recreational concerns in forest management and planning. Half of the Atlantic countries (Iceland, Belgium and the UK) are (strongly) dissatisfied with the present education level. Denmark is happy with the present level of education, while Ireland and the Netherlands are neutral. All countries are of the opinion that working experience provides valuable knowledge of recreational issues and management techniques. The Netherlands thinks that recreational concerns are very well considered through working experience.

There are no regional differences in foresters' recreational education and working experience except in the UK. Forest and recreation managers in the densely populated south of England have more experience in managing the issues and conflicts of recreational use than many of their counterparts in more remote areas of the UK where these issues are less common. In all countries, there are differences in foresters' recreational education and working experience between different types of owners. In Iceland, forestry education was not offered, meaning that the Icelandic foresters received their education abroad. However this is changing as the newly established agricultural university now offers forestry education at the university level. Training courses can also be offered to practicing foresters. In Ireland the Tree Council sponsored a training course for anyone working in outdoor recreation planning and design which was held annually for many years. Specific aspects are also provided by the Forestry Training Services unit of the British Forestry Commission. These help to update foresters' knowledge over time.

Each Atlantic country has government institutions concerned with forestry issues. These forest administrators play a similar role in Belgium, Denmark, Iceland and Netherlands. In all of these countries, these institutions have the task of planning facilities for forest recreation. Except for Belgium, the forest administrations of all countries are involved – although not equally intensively – in designing the monitoring framework for recreational issues as well as the monitoring process itself. Monitoring is especially important for the British Forestry Commission. Forest managers in the Atlantic countries all have similar tasks regarding forest recreational issues. All emphasized the importance of planning, information sharing, environmental education, visitor guidance, and establishing recreational infrastructures as some of their key tasks.

2.5 Conflicts

2.5.1 Conflicts: Functions, Competence and Scale

Recreation in forests is just one of many forest functions. Brandt and Vejre (2003) propose three different dimensions of functionality with respect to landscapes that also apply to forests. The first dimension focuses on pure ecosystem functionality – dealing with fluxes of energy, matter, and species. The second dimension is concerned with human-nature interactions resulting in different patterns of land use. What these two approaches have in common is that they deal with "material functions and their spatial intensity". The last dimension is called "transcending functionality" and includes not only non-material functions but also values and feelings about forests. Following this concept, recreation can be represented in both the "land use" dimension (e.g. hiking on a trail) and in the "transcending" dimension (e.g. the view of a forest or the cultural value of its existence). Often the two functions are closely interrelated, but not always. Hence, awareness of the latter two dimensions of functionality is important when considering the regulation of forestry and forest recreation (including conflict mitigation).

Conflicts can occur when a recreational function is, to some degree, incompatible with other forest functions and is perceived to be a problem among different user groups[3]. The conflicts can also exist between different types of recreational functions in terms of both conflicting recreational "land uses" (e.g. hiking versus mountain biking on the same trails) and a recreational "land use" that conflicts with a "transcending" function (e.g. snow-mobile use versus experiencing the quietness). In these examples, the multifunctional use of forests causes the conflicts. The main reason for conflicts is not limited to functions. Conflicts can also arise between two or more visitors engaged in the same activity, for example when two persons go to the same spot in a forest to find quietness and solitude ("crowding"). Consequently three different types of conflicts can be distinguished among functions of both physical (land use) and mental (transcending) origin:

[3]There are a lot of functions that are (partly) incompatible on a certain location – e.g. draining wet areas and having beavers in the same area – but if no human actors consider the incompability as a problem, is it not regarded as a conflict here. The situation might be considered as a problem by future generations and therefore relevant to consider in relation to sustainability.

- forest recreation functions versus other forest functions;
- recreation function A versus recreation function B;
- recreation function (person 1) versus same function (person 2).

Different spatial competence among actor groups can also cause conflict[4]. The classic example is the territorial competence of a private land owner versus a forest visitor with limited competence who only has access on forest trails by foot. Often, the distribution of competence between users is found in the law, but voluntary or commercial "outsourcing" of competence from the owner to visitors is also possible. Access by local people to forest areas with legally limited access or the renting out of the game hunting privileges are examples of outsourcing. The configuration of competence can vary between types of forest ownership (private, community, state etc.), from actor to actor, and from one country to another because of differing laws or customs. There are two levels of conflicts caused by the distribution of competence:

- Disagreement on competence. This occurs when one of the actors disagrees with the existing distribution of competence. This can be local people, who are dissatisfied by the lack of access to a private forest, even if the landowner has the legal right to deny access. It can also be a forest owner who does not allow people access to his/her forest, although the law obliges him/her to do so. Conflicts can be expressed as local arguments, poor neighbour relations, or policy discussion, but have no material consequences for the forest.
- Exceeding competence. The second level is when an actor exceeds his/her competence; a visitor goes horseback riding or an owner puts up "access forbidden" signs, even if both are illegal.

It is probably seldom that conflicts are solely derived from different competences, e.g. between landowners and visitors. Rather, conflicts tend to have specific reasons. To return to the landowner versus visitor conflict: the landowner perceives a conflict between, for example, his/her hunting activities and the visitors' picnicking activity. In other words the actors' perception of the incompatibility of functions is the main reason for the conflict. However the different competences (or power relations) are important components for understanding conflicts.

Conflicts depend on time and space. Some functions are not usually simultaneously carried out and may not result in conflicts if they are separated over time. Recreational activities can conflict with forestry operations while the forest owner is actually cutting trees, but not necessarily outside this period. Conflicts are also related to spatial distribution of the forests (area and location) and to demands for recreational activities (variety of activities and numbers of persons). Potential conflicts may never be realized because of adequate space or designation of areas for different functions.

---

[4]Competence is here understood as the power/right to enforce a certain spatial practice or state, e.g. landowners have competence to decide on land use within the legal frames that has been given by the authorities due to their spatial competences. See Hägerstrand (1995) for further discussion about competences and space.

2.5.2 Types of Conflicts

Five major conflicts recorded in the questionnaire include all three functional types discussed above. Each conflict can be ranked on a scale ranging from not important at all (1) to very important (5). Conflicts become especially apparent in situations with high visitor pressure and low forest supply. As the situation in the Atlantic countries with regard to these aspects is quite varied, the importance of each conflict has been divided by the amount of forest and wooded area per inhabitant (FOWL). Figure 2.6 shows the results or conflict density for each country.

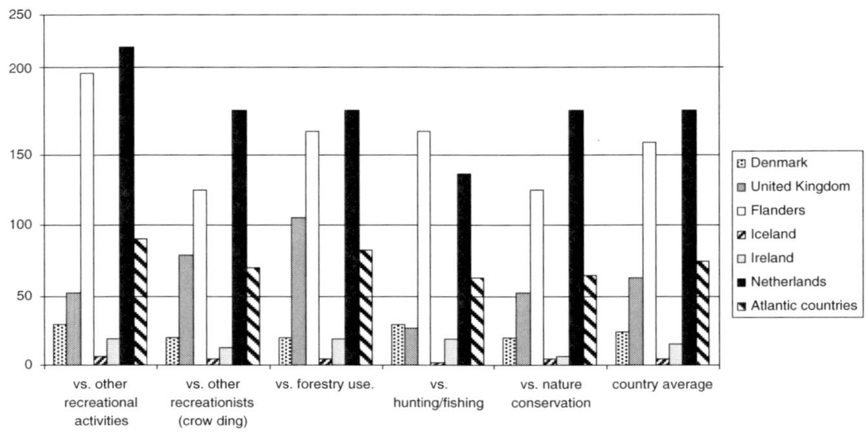

Fig. 2.6 Conflict density (conflict importance/FOWL)

From this graph it can be seen that there is a large variation in the conflict scores. The two extremes are: the Netherlands showing strong conflicts on "other recreational activities" (score = 228) and Iceland showing almost no conflicts between "fishing and hunting".

For the Atlantic countries as a group, the importance between the five types of conflicts does not differ strongly from one to another. However, conflicts between different recreational activities are considered to be most important. Conflicts between recreation and hunting/fishing are the least important. Conflicts resulting from forestry use, crowding, and nature conservation are in the middle of the spectrum.

At the country level, there is a distinction in three levels of conflict occurrence. First of all, conflicts overall seem to be most important in the Netherlands. This would be expected as this country is very densely populated and has a low forest cover. The UK and Belgium hold an intermediate position. Belgium, however, is not regionally uniform in its conflict level. It must be observed that the conflict level for each type of conflict of Flanders is similar to that of the Netherlands. Although the FOWL in the UK is not that much higher than that of the Netherlands, the state forest sector of the UK believes this is attributable to their organisation's careful planning of how areas are zoned and managed specifically to minimise conflicts. The Dutch and the Flemish State Forestry Service claims to do the same, however their conflict rating is considerably higher. Lastly, the least degree of conflict intensity is experienced by Denmark, Ireland, and Iceland. In keeping with this pattern, Iceland shows the lowest level of conflict and has the highest FOWL per capita. This indicates that although zoning can prevent conflict, when the use of forests becomes intensive and highly diverse, conflicts are inevitable.

Similarities and differences between the types of conflict indicate that all countries, except for the UK, consider conflicts between recreational activities to be the most important. Especially conflicts between motorized and non-motorized activities, between walkers and bikers, and between hikers with dogs not on leads and other recreationists.

In the UK conflicts between recreation and forestry operations are considered the most important. With a small land area and high population density, almost all forest operations inevitably have significant impacts on the public. Consequently, it is not surprising that the Dutch and Belgium experts also rate this type of conflict highly. These three countries experience the most problems with crowding. Crowding problems, of course, are most extreme in the Netherland and Flanders. The seriousness of these conflicts is, however, not as great as would be expected. Inhabitants of these regions have become accustomed to high visitor density in the forests and other natural areas. Belgium, Denmark and Ireland have relatively many problems with hunting and fishing activities in relation to other recreational activities. Recreation-nature conservation conflicts are present in both the Netherlands and Belgium. Both countries are trying to achieve high levels of biodiversity in their forests and nature areas. This is quite often in contradiction with recreational goals.

2.5.3 Important Activities and Their Conflicts with Nature Interests

Because of climate and topography, winter sport activities such as skiing are rare in the Atlantic countries. Only Iceland and Belgium (Wallonia) report these activities and both their importance and conflict levels are low (see Table 2.3). Only cross-country skiing in Wallonia causes some damage to nature areas.

From the second group of activities, hiking is the most common but rarely a conflicting activity. Only Iceland reports some conflicts. Orienteering in Belgium and the UK and geo-caching in the Netherlands can cause environmental problems, most probably related to its "free strolling" aspect that does not take into consideration environmentally sensitive spots. Next, Belgium, the UK and the Netherlands report high importance levels and a conflicting element of "hiking with dogs not on leads". Chasing dogs can disturb wildlife and birds. Farmers (who are often private woodland owners) are sometimes seriously bothered by dogs not on leads – though this has more to do with the dog bothering farm animals.

Table 2.3 Importance and frequency of conflicts between recreational activities and natural environment/conservation (importance: 1=low, 3= high, frequency of conflicts: 1= rarely, 3=frequent)

	Importance of activity							Frequency of conflict						
Activities	BE	DK	IC	IR	NL	UK	av.	BE	DK	IC	IR	NL	UK	av.
Cross-country skiing	1	–	1	–	–	–	1.5	2	–	1	–	–	–	1.5
Alpine skiing	1	–	1	–	–	–	1	1	–	1	–	–	–	1
Ski touring	2	–	1	–	–	–	1	1	–	1	–	–	–	1
Snowshoeing	–	–	–	–	–	–	–	–	–	–	–	–	–	–
Snowmobiling	–	–	1	–	–	–	1	–	–	1	–	–	–	1
Orienteering	2	2	1	1	1	1	1.3	2	1	1	1	1	2	1.3
Nordic/power walking	1	1	2	–	1	–	1.3	1	1	2	–	1	–	1.3
Geo-caching	1	–	–	–	2	1	1.3	1	–	–	–	2	1	1.3
Climbing	2	–	1	1	–	1	1.3	2	–	1	1	–	1	1.3
Hiking	1	3	3	1	3	1	2	1	1	2	1	1	1	1.2
Mountain biking	2	2	1	2	2	1	1.7	2	2	1	2	2	2	1.8
Motorized sport	3	1	1	3	2	1	1.8	2	1	1	3	3	1	1.8
Horse riding	2	2	2	2	2	1	1.8	2	2	2	1	2	1	1.7
Cycling	1	3	2	1	3	1	1.8	1	1	1	1	3	1	1.2
Horse carriage driving	1	1	1	–	1	1	1	1	1	1	–	1	1	1
Camp fires	2	2	1	1	1	1	1.3	2	2	1	1	1	1	1.3
Camping	2	1	2	1	2	1	1.5	2	1	2	1	1	1	1.3
Collecting berries, mush-rooms, minerals, etc.	2	2	3	1	1	2	1.8	2	1	1	1	1	1	1.2
Wildlife/bird watching	2	3	2	1	2	1	1.8	2	1	1	1	1	1	1.2
Picnicking	1	3	3	1	2	1	1.8	1	1	1	1	1	1	1
Motorized water sport	1	2	1	–	1	–	1.3	1	3	1	–	2	–	1.8
Canoeing	2	2	1	1	1	–	1.4	2	2	1	1	1	–	1.4
Fishing	1	2	2	3	2	1	1.8	1	1	1	2	1	1	1.2
Swimming	1	3	1	–	1	–	1.5	1	1	1	–	1	–	1
Country average	1.5	1.7	1.5	1.4	2.1	1.1	–	1.5	1.4	1.2	1.3	1.4	1.1	

The third group of activities, i.e. cycling, mountain biking, horseback riding and motorized sport, are quite common in all countries (horse carriage driving is not reported in Ireland). The highest conflicting activities (mean value $\geq$ 1.8) in the respective countries are motorized sport, although in most countries forbidden, and mountain biking. Motorized sport disturbs people, wildlife and nature with its noise, air pollution and erosion, causing it to be rated by the country experts of Ireland, Wallonia and the Netherlands as the most highly conflicting activity.

Sedentary activities, such as collecting berries, building camp fires, camping, picnicking and wildlife/bird watching, are seldom reported as conflicting.

In general, the most densely populated countries, i.e. the Netherlands, Belgium and Denmark, score highest on the importance of all recreational activities as well as having the greatest frequency of conflicts between these activities and nature conservation and environmental concerns. The UK, however, is an exception, having both the smallest frequency of activities and conflicts. However, factors other than user density may influence the level of conflicts – including the level of recreational planning and the use of zoning of incompatible forestry uses and recreational activities through time and space to reduce conflicts (see Section 6).

2.5.4 Impacts and Disturbance of Other Societal Activities on Recreation

The experience of nature can be highly disturbed by all kinds of intruding activities, such as noise of traffic and aeroplanes or the visual impact of forest exploitation or tourism development. In all countries, recreational experiences are negatively influenced to some extent. Iceland is a positive exception: none of the mentioned disturbance factors (see Table 2.4) causes any conflict. Consequently, Iceland is excluded from this analysis.

The most important disturbances which cause conflicts (both mean value $\geq$1.5), are rubbish dumping, traffic noise, road construction, and large clear cuts. The first, the dumping of rubbish ranging from small recreational waste such as sweet wrappers and tins, to bulky household items such as furniture and to old cars and rubbish bags, occurs in all Atlantic countries. It is especially a problem in Ireland, the Netherlands and the UK.

In the Netherlands traffic noise and road construction are also found to be very disturbing. Traffic noise is not disturbing in Ireland; road construction causes little disturbance in Denmark and the UK. Large clear cuts can cause problems in Belgium, Ireland, and the Netherlands. Although there are no large clear cuts in Flanders and the Netherlands, it should be kept in mind that forest visitors are very sensitive to timber cutting in general. In Wallonia, most forests are regenerated by clear cuts ("futaie régulière par parquet"). Clear cuts can potentially cause conflicts especially when they are too big, but in some cases people enjoy them as they open the landscape and offer new panoramas. Aircraft noise is important in all countries,

Table 2.4 Importance and frequency of disturbance factors on forest based recreation and nature based tourism (importance: 1=low, 3= high, frequency of conflicts: 1= rarely, 3=frequent)

	Importance of activity							Frequency of conflict						
Disturbance factor	BE	DK	IC	IR	NL	UK	av.	BE	DK	IC	IR	NL	UK	av.
Rubbish dumping	2	2	1	3	3	2	2.2	2	2	1	3	3	3	2.3
Traffic noise	2	2	2	1	3	2	2.0	2	2	1	1	3	2	1.8
Road construction	2	1	1	2	2	1	1.5	2	1	1	2	3	1	1.7
Large clear cuts	2	2	1	2	1	1	1.5	2	1	1	2	2	1	1.5
Air craft noise	2	1	2	1	3	2	1.8	2	1	1	1	2	1	1.3
Military noise	1	1	1	1	2	1	1.2	1	1	1	1	2	1	1.2
Wind turbines	1	1	1	1	2	1	1.2	1	1	1	1	2	1	1.2
Exploitation of natural resources	2	1	1	2	2	1	1.5	2	1	1	1	1	1	1.2
Industrial noise	1	1	1	1	1	1	1	1	1	1	1	1	1	1
Other emissions (odour, dust)	1	1	2	1	1	1	1.2	1	1	1	1	1	1	1
Visual problems of damaged forests (i.e. insects, fires)	2	1	1	1	1	1	1.2	1	1	1	1	1	1	1
Fire breaks	1	1	1	1	1	1	1	1	1	1	1	1	1	1
Infrastructure for tourism (ski-lifts, cable cars, etc.)	1	1	1	–	2	1	1.2	1	1	1	–	1	1	1
Country average	1.5	1.2	1.2	1.4	1.8	1.2	–	1.5	1.2	1	1.3	1.8	1.2	–

except for Denmark and Ireland, but cause the greatest conflicts in Belgium and the Netherlands. Forest visitors in the Netherlands also experience conflicts from military noise and wind turbines. The exploitation of other natural resources affects forest visitors in Belgium. The Atlantic countries experience lesser effects of other disturbance factors such as industrial noise, emissions (odour, dust), visual problems from damaged forests (i.e. from insects, or fires), fire breaks, and tourist infrastructure (ski-lifts, cable cars, etc.). However, electricity transmission lines sometimes affect forest landscape and reduce attractiveness. Wallonia mentions problems with quarries, sites where rocks are extracted, in the forests. Most of them are located in tourist valleys (Meuse and Ourthe), causing disturbances such as noise and dust and degradation of the landscape quality.

Again the average values for each country of both the importance and the conflicting nature of disturbance factors indicates that the most densely populated countries, i.e. the Netherlands and Belgium, have the highest overall score.

2.6 Planning, Management and Monitoring

2.6.1 Planning for Recreation and Nature Tourism

For all countries, except Iceland some form of independent planning for recreation and nature tourism is undertaken at regular intervals or in special cases (Table 2.5). Planning is undertaken regularly in Flanders, Netherlands and the UK as a routine part of integrated forest management planning systems. In Denmark, Ireland and Wallonia this planning takes place only in special cases, when e.g. new types of activities or facilities in the state forest are being considered or when forests are situated close to urban or tourist centres.

Table 2.5 Availability of independent planning, either legally required or optional

	IC	IR	DK	UK	BE	NL
Independent planning for recreation and nature tourism in forests	No*	Yes, in special cases	Yes, in special cases	Yes, at regular intervals	Fl: yes, at regular intervals W: yes, in special cases	Yes, at regular intervals
Legal or optional requirement for forest owners	–	–	Optional	Optional	Both legally required and optional (W)	Optional

* Although planning is not mandated by law, recreational planning is available when asked for. The Iceland Forest Service has done it for some areas of the national forests; they will also work with municipalities to help them in planning recreational areas.

The report responses do not clearly indicate the reasons why forest recreation planning is undertaken regularly or only in special cases in the Atlantic countries. Possibly different countries are describing similar planning systems in different ways. By way of example, in the UK, recreation planning is being undertaken in a variety of different ways depending on the situation:

- Recreation planning per se is not a legal requirement for forest owners. Planning/zoning of woodland management for recreational activities is undertaken every 10 years under the forest planning system;
- Detailed plans for a "change of use" from forestry require statutory planning permission for the provision of any man-made infrastructure (tracks, car park, toilets) for all forest owners and managers;
- Regionally, the emphasis on recreation planning is quite different depending on factors such as population density/distribution, provision of other recreational opportunities, and the understanding of why forests exist (timber, rural employment, recreation, etc.);
- Private forest owners may employ independent consultants to develop recreation schemes, but due to a general reluctance to provide recreation on private land (risk of liability) the need for recreation planning is less. Private forest owners in the UK must demonstrate some degree of recreation planning before being eligible for certain additional grants.

The planning system of state owned forests in the Netherlands is similar to the situation in the UK: every 10 years new (recreation) planning is required and the quality depends on the local situation and the competency of the involved managers. The same holds true for state owned forests of Denmark (time perspective is 15 years) and Flanders. The Dutch situation for private owners is similar to the British context. In Flanders, private forest owners are encouraged to make a recreation plan every 20 years as part of the forest planning system. Since 2003 subsidies for recreation (accessibility) are provided for private owners if accessibility criteria are met: 100 euros/ha/yr for play areas with free access for youth movements (and others) and 50 euros/ha/yr for maintaining accessibility for hikers on the designated pathways. In the other Belgium region, Wallonia, nothing like this exists: planning for forest recreation is basically optional for private forest owners.

Recently, the Dutch national government has changed its conventional recreation planning in terms of goals, actions, etc. into a planning system of required outputs. This Management Programme is meant to subsidize nature and forest owners in a controlled way. It is an output-orientated endowment scheme for payment for output of specified forestry goods and services. It places less emphasis on the production function and more on landscape and recreation functions of forests. Subsidies for recreation are provided if accessibility criteria are met (Hoogstra and Van Blitterswijk 2002).

In Ireland recreation is re-emerging as an important forest planning issue from a period where industrial commercial forestry was the primary objective. This is in response to increased urbanisation and an increasingly negative perception of the forest industry. Recreation provision is dealt with at a local (forest district) level with support from a small national team. In some recreational areas statutory planning permissions and licences maybe required.

In Denmark, the Netherlands and the UK planning for forest recreation is usually an optional requirement for forest owners, though for certain aspects of recreation, development planning is a legal requirement (as explained above). For state owned and other public forests in Flanders, recreation planning is on principle a legal requirement. Managers of privately owned forests are encouraged to make a recreation plan through the provision of subsidies. For state owned forests of the Netherlands recreation planning is also legally required.

2.6.2 Guidelines and Planning Steps

Guidelines for planning have been produced in Denmark, Ireland, the Netherlands, and the UK in a variety of formats across the countries (nature guides, recreation policy, planning and control systems and design guides). Only in Iceland and Belgium are guidelines unavailable. Only the Netherlands mention specific planning tools and research. Only the UK mentioned development of alternative formats (design guides, specifications, web based information, some independent training programmes and in the state sector guidance and management between recreation managers and in house training).

The responses indicate that in the Netherlands and the UK a wide range of steps are undertaken for inventory, analysis, goal setting and monitoring at regular intervals during the forest recreation planning process (Table 2.6). The UK and the Netherlands are two of the countries that state independent planning for recreation and nature tourism is undertaken at regular intervals. This correlation indicates that where recreation planning is undertaken at regular intervals, more complete systems were developed to aid the process.

Table 2.6 Steps that are undertaken for planning forest recreation at regular intervals

Steps in planning forest recreation	Countries*
Inventory of ecological conditions	BE, UK, NL
Inventory of social conditions	UK, NL
Analysis of conflicts	UK, NL
Definition of vision and goals	BE, DK, UK
Participation in the definition of goals	DK, UK, IC, IR, NL
Definition of standards	UK, IR, NL
Deduction of measures	NL
Participation in the deduction of measures	NL
Monitoring	UK, NL
Other, local consultation	IR

* Iceland has no planning tradition with regards to forest recreation and nature tourism

Survey data from national surveys, regional surveys, forest administration data and from other organisations can aid in the planning and management of recreational areas. Denmark and the UK report having the most survey data to use in planning. The Netherlands and Belgium have some data, Iceland has local data and Ireland does not report having any data. This may reflect different traditions, types and stages of conflict management with some countries focusing more on planning and regulations and others on the documentation of activities and related problems. It is likely, that research activities increase as conflicts increase. Several Atlantic experts have indicated that new studies regarding tourism and recreation are presently carried out.

2.6.3 Public Participation

Public participation in the recreation planning process is undertaken as a matter of course in all countries but Belgium. Flanders has displayed a great commitment to forest recreation by establishing a legal requirement for forest recreation planning; yet so far this process has not involved public participation. Recently, however, Flanders is experimenting with public participation by means of developing new accessibility rules jointly with private forest owners and recreationists. Interestingly, and against the trend, this is an aspect well covered by Iceland. This is due to the fact that the majority of recreational forests surrounding urban areas are managed by the local forest societies. These societies are local NGOs with public membership and well-organised public participation in their decision-making and negotiations with the local municipalities, who usually own the land. Table 2.7 identifies the range of groups involved in the public participation process.

Inevitably, those individuals or groups that can be identified are being consulted by the majority of countries with the exception of Flanders. Those groups (e.g. berry pickers) who do not have formal local representation are missed during the consultation by all countries except the UK. There, temporary notices are placed at

Table 2.7 Groups that take part regularly in public participation in the definition of visions and goals

Groups in public participation processes	Countries*
Land owners	DK, UK, IC, NL, IR
Land users	DK, UK, IC
Local residents/other citizens	DK, UK, IC, IR, NL
Hiking clubs and sport associations	DK, UK, IC, IR, NL
Mushroom/berry collectors	UK
Hunting and fishing representatives	DK, IC, IR
Community representatives	DK, UK, IC, IR, NL
Nature associations	DK, UK, IC, IR, NL

* Belgium only has "ad hoc" public participation

woodland entrances inviting such groups to attend public consultations. In Ireland there is considerable local consultation on recreation developments as part of the overall sustainable forest management process.

2.6.4 Monitoring

Recreation monitoring is the systematic collection of data, such as visitor numbers, level of outdoor facilities and visitor satisfaction. It is used to define and evaluate specific recreational goals. Monitoring is carried out as necessary, on a regular or irregular basis, by most Atlantic countries. Only in the UK, Ireland and the Netherlands is there a regular monitoring process. In the UK survey and monitoring is undertaken by a specific department within the Forestry Commission specialising in statistics. Nationally collected data is compiled in a report and published every 1–2 years. The Forestry Commission also has a Research agency that in the last few years has been developing research projects in the fields of recreation and social science. Historically, Forestry Commission work focused on timber production, the new establishment of this special department reflects an increased awareness of the importance of social aspects of the forest. In Ireland, monitoring is part of the Sustainable Forestry Management (SFM) initiative. The state forestry company, Coillte, has completed an evaluation of the contribution of recreation to economic activity and national well-being (Fitzpatrick Associates Economic Consultants, 2005). Monitoring in the Netherlands takes place via the output-oriented function endowment scheme that was discussed earlier. The endowment scheme provides subsidies for nature conservation, landscape, and recreation if certain criteria are met.

Each country was asked to evaluate the integration of forest based recreation and nature tourism in the planning and monitoring of the country's forests. Denmark and the UK recorded slightly higher than average scorings, Ireland and Netherlands recorded satisfactory scores, whilst Belgium and Iceland registered lower than satisfactory ratings. This response indicates that forest recreation planning (and monitoring) tools are in the early days of development as forests are only recently regarded as important for recreation opportunities in the Atlantic countries, as opposed to exclusively timber production areas.

The lack of planning and monitoring (tools) for forest recreation may be explained by the recent rise of forest recreation and nature tourism. The effects of these increases have only been experienced during the last decade(s). These are the results of increasing urbanisation and population numbers. In the past, the level of use was so insignificant that planning and monitoring of recreation seemed to be irrelevant. Forest managers simply added recreation to their daily duties. At present, the demand for recreation increases in both quantity and quality (more services, more diversity in supply, more information and education). This requires more knowledge, more educational training and more sophisticated planning and monitoring systems.

2.7 Recreational Infrastructure

2.7.1 Supply of Basic and Additional Infrastructure

Within this section, the basic infrastructure for recreation, such as forests roads and trails, and other infrastructure elements, such as benches and nature trails, will be evaluated for urban areas, rural areas, and remote areas. Urban areas refer to cities, towns and residential areas, rural refers to the countryside in which agriculture plays an important role, and finally, remote refers to sparsely populated areas that contain non-cultivated areas or wilderness areas. It must be understood that the interpretation of these concepts varies significantly throughout the Atlantic countries. It is important to notice that Denmark and the Netherlands state that they do not have any remote areas in their countries. The Belgium remote areas are only situated in Wallonia. Evaluation scores vary from "unsatisfactory" to "very good". Of course, the evaluation takes differing visitor expectations concerning the three geographical areas into account: in remote areas, limited infrastructure does not necessarily mean that it is unsatisfactory, because people visiting these areas are likely to be primarily interested in experiencing wilderness and a lack of infrastructure. The reverse is true for urban areas.

As can be seen in Fig. 2.7, the supply of basic infrastructure in urban areas and rural areas is described in most countries as (very) good. Although the Irish situation is good in rural and remote areas, the quality of basic infrastructure in urban areas is considered to be poor. The reverse is true for the UK: the urban forest infrastructure quality is good, but the rural and remote forest infrastructure is less satisfactory. The urban forest infrastructure in the UK is reasonably well funded and relatively new. This new funding tends to follow political agendas and therefore focuses on high population areas. This is a relatively new trend. Iceland is an exception in this context. There is in general an unsatisfactory level of basic infrastructure, especially in rural and remote areas.

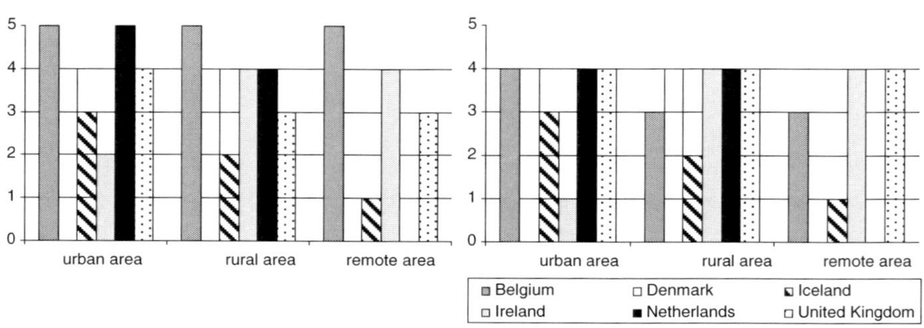

Fig. 2.7 Supply of basic infrastructure (*left figure*, **a**) and additional infrastructure (*right figure*, **b**) (no bar = non-existent, 1=unsatisfactory, 5=very good)

The country representatives were also asked to assess the additional infrastructure (see Fig. 2.7b). Half of the countries (Denmark, the Netherlands and the UK) evaluated the provision of additional infrastructure in all areas as good. The situation in Belgian urban areas is well evaluated; the provision of facilities is reasonable in rural and remote areas. The situation in Iceland, again, is considered less satisfactory. This is however less problematic than the Irish situation, which is considered very unsatisfactory. Urban forests in Ireland are neither plentiful nor well developed. Of the countries with remote areas, the UK is satisfied with the present supply (and new investment is increasingly hard to come by), whereas Iceland is dissatisfied with the amount of available forest land in these areas. In Iceland there is a demand for forest recreational areas in remote areas, especially during the intensive tourism season from May to September. Therefore, more investment to increase the supply of both basic and more advanced infrastructures is needed in those areas.

2.7.2 Additional Facilities

Table 2.8 presents the estimated occurrence of additional facilities in forest areas for all six countries. It can be concluded that in general the more urbanized an area is, the more infrastructure is available. Exceptions are huts, toilets, kiosks, camping grounds and game preserves. These facilities are most frequent in rural areas.

In urban areas, the most widely distributed facilities (minimal judgement of distribution is "reasonable") are benches, picnic areas, nature/sport trails and rubbish bins. Danish and Dutch forests especially offer recreational facilities, while the British and Icelandic forests somewhat less. The Belgian forests, though, are less equipped. Urban forests in Ireland have hardly any additional facilities for visitors.

Table 2.8 Occurrence of additional infrastructure in urban, rural and remote areas (1= rarely, 2=reasonable, 3=frequently)

Facilities	Urban areas	Rural areas	Remote areas (not in DK and NL)
Benches	2.5	2.2	1.3
Picnic areas	2.3	2.2	1.3
Nature/sport trails	2	2	1.5
Rubbish bins	2	1.5	1
Playgrounds	1.8	1.7	1
Environmental education infrastructure	1.7	1.5	1.3
Fire places	1.7	1.2	1
Look outs/viewing platforms	1.6	1.5	1
Toilets	1.3	1.5	1
Huts	1.2	1.3	1.3
Kiosks	1.2	1.6	1
Camping grounds	1	1.7	1.3
Game preserves	1	1.4	1.3

Atlantic rural forests offer less than urban forests. The following facilities are (almost) reasonably well distributed: benches, picnic areas, nature/sport trails, playgrounds, and campgrounds. Again, Belgium and Ireland have a limited range. The provision of picnic areas in Ireland is very good; recreational use has traditionally been comprised of walking and picnicking, rather than for instance, camping or fire lighting common in other European countries. New recreational activities are now being developed and encouraged in Irish forests. Remote areas of Belgium, Iceland, Ireland, and the UK have hardly any additional infrastructure. Only Belgium and Ireland indicate that environmental education infrastructure, such as nature trails, is reasonably available.

Most countries do not feel that forests have been over-equipped for experiencing nature in recent years. Only the Netherlands has a neutral attitude regarding this statement.

2.7.3 *Quality of Infrastructure*

Finally, the quality of the recreational infrastructure is also important (Fig. 2.8). Quality refers to good original design, regular renovations, good maintenance, checking of safety features, and evolution of design over time. Only in the Netherlands is the quality of infrastructure in both urban and rural areas considered to be excellent. The other country experts have a neutral attitude (not excellent, not unsatisfactory) towards the level of quality of recreational infrastructure in all areas. Only in Iceland is the provision in both rural and remote areas regarded as unsatisfactory. Where the UK is of the opinion that the quality is urban areas is excellent, Ireland thinks it is unsatisfactory.

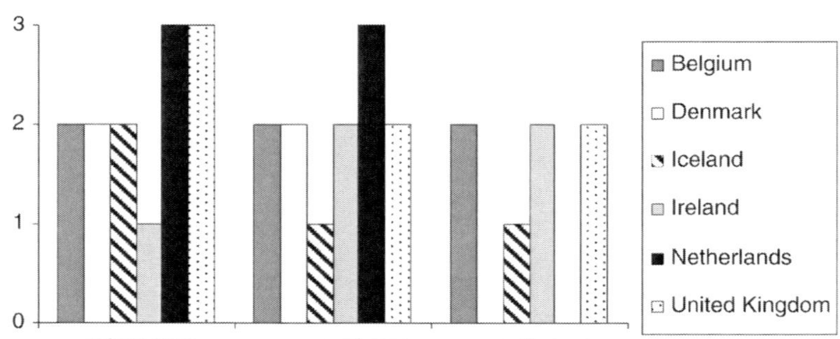

Fig. 2.8 Quality of infrastructure for recreation and nature-based tourism (no bar = non-existent, 1=unsatisfactory, 3=excellent)

2.8 Approaches to Visitor Management

In order to maintain and develop recreational infrastructure and services needed to facilitate appropriate visitor behaviour and comfort, different approaches to visitor management have been applied. Obviously, each of these approaches aims at the reduction of specific (environmental, social or economic) conflicts occurring in recreational settings. Although the variation in approaches to visitor management is not a major factor for dividing the different regions in Europe, some particularities and trends are observable in each region.

In this section special attention is paid to the use of management actions aimed at visitors in the Atlantic zone. The frequency with which such actions are used and their acceptance by the public will also be examined. Various management actions are classified into strictly restrictive actions (prohibitions), soft actions (infrastructure, sign posting) and facilitating actions (guiding, education) (Table 2.9).

Table 2.9 Occurrence of management actions in urban, rural and remote areas (1= rarely, 2=regularly, 3=frequently) ordered by degree of acceptance by the public (1=high, 2=medium, 3=low

Management actions	Urban areas	Rural areas	Remote areas (only IC, IR, BEL)	Acceptance by the public
Restrictive actions				
Rules and regulations	2.3	2.2	1.7	1.7
Barriers and fences	2.2	2	2.3	2
Prohibitions	2	1.8	1.3	2
Marking protected areas	1.5	1.5	1.3	2
Limiting visitors' entrance	1	1	1	2.2
Entrance fee	1	1.4	1	2.4
Soft actions				
Improving areas with new attractive infrastructure (e.g. attractive roads, view points, play grounds)	2.2	2	1.5	1.2
Signposting for visitors to act in a responsible way	2	2	1.7	1.2
New infrastructure for ecological education	1.7	1.7	1.3	1.2
Facilitating actions				
Guiding visitors by ranger system	1.7	1.5	1	1.4
Guiding visitors by thematic tours	1.7	1.8	1.3	1.6
Providing special training for user groups (e.g. climbers)	1.2	1	1	1.7
Providing special training for commercial providers (e.g. boat or ski-rental)	1	1.2	1	2

Generally throughout Europe it is usual to implement measures for the guidance of visitors in forests and nature areas. It may be useful to evaluate these measures according the degree of urbanisation, rurality or remoteness. Yet, because of the common occurrence of legislation regarding recreation throughout the different regions, they are also evaluated according to their "national" backgrounds.

2.8.1 Restrictive Management Actions

First, the intensity to which visitors are imposed upon by restrictive actions in recreational use was evaluated. It can be observed that rules and regulations, barriers and fences and prohibitions are widely applied, especially in urban and rural areas. Iceland and Ireland make frequently use of barriers and fences in their remote areas. No clear tendencies concerning this matter are observed throughout the Atlantic region. In Denmark and Iceland, prohibitions are rarely used for visitor management. By contrast, restrictive actions are applied more frequently in the Netherlands, Belgium, and to a lesser degree in the UK and Ireland. A feasible explanation for the observed differences may stem from the high population density and the degree of urbanisation in these countries. Because recreational pressure is higher in the vicinity of densely urbanised areas, natural resources are more frequently exposed to impacts and crowding conflicts between user groups occur more often. For this reason, it is likely that these countries have incorporated more prohibitions into their legislative frameworks. There may also be management traditions that vary between countries. Denmark and Iceland are more closely related in their approaches than are the other countries in the Nordic region (e.g. Norway, Sweden). The collection of entrance fees and the limitation of visitor accessibility are rarely applied.

2.8.2 Soft Management Actions

Soft actions, such as signposting and the development of attractive infrastructure, are widely implemented in the Atlantic region. Signposting is considered to be an especially important aspect of visitor guidance, in particular recreational routes for varying user groups (e.g. hikers, mountain bikers, horse riders). Ecological education facilities are also relatively frequently used. All soft actions are often placed in easily accessible, urban and rural areas to improve recreational experiences.

2.8.3 Facilitating Management Actions

Throughout the Atlantic region various types of ranger systems, i.e. park stewards with some responsibilities for recreation management, are implemented to enhance services to visitors. These systems may be related to the provision of thematic tours for visitors in a natural environment, organised on a regular basis. This is common in the Netherlands. In Denmark guided activities are increasingly being used. Special training for user groups or commercial providers in the recreation sector occurs very rarely in the Atlantic region. Presumably, training is organised more often by private initiatives and specialised organisations.

2.8.4 Urban–Rural Continuum

Overall, it can be concluded that the frequency of management actions implemented is highest in urban areas and decreases with increasing rurality. The differences between urban and rural areas, however, are small. This pattern can also be observed for each group of actions, i.e. restrictive actions, soft actions and facilitating actions. The Netherlands and the UK most frequently use all types of management actions. Both Denmark and Iceland apply the least amount of visitor management. Belgium and Ireland are in the middle. The region of Flanders resembles the Netherlands most, whereas Wallonia is similar to Ireland.

2.8.5 Acceptance of Management Actions by the Public

The acceptability of the different management actions in the Atlantic region is illustrated in Fig. 2.9 (see also Table 2.10). This shows that the experts are of the opinion that the soft actions and visitor guiding measures (thematic tours), are considered to be widely accepted by the public. Restrictive actions in general have little public support, although "rules and regulations" are understood as being necessary. There is a rather low acceptance of restrictive actions in Iceland and Ireland.

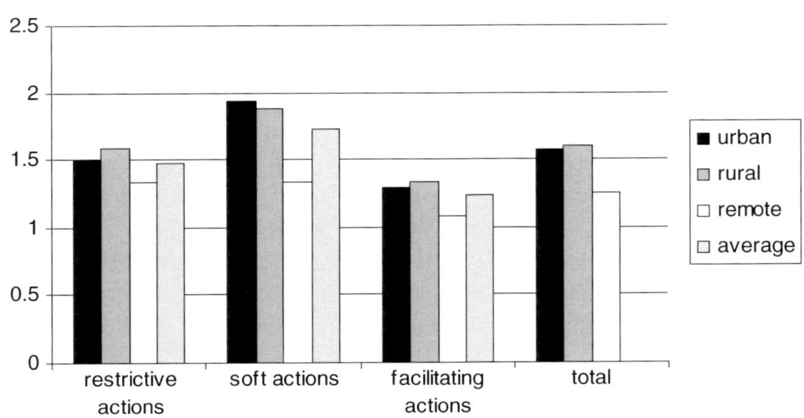

Fig. 2.9 Occurrence of management actions in the Atlantic region, in general and in urban, rural and remote areas (0=non-existent, 1= rarely, 2=regularly, 3=frequently) (remote only applies to Iceland and Ireland)

Table 2.10 The level of acceptance in the Atlantic region for the various management actions (high >= 2.5, medium > 1.5 and < 2.5, low =<1.5)

Acceptance by the public	Restrictive management actions	Soft management actions	Facilitating management actions
High		Attractive infrastructure Educational infrastructure Signposting responsible visitor behaviour	Guiding ranger systems
Medium	Rules and regulations Prohibitions Marking protected areas Barriers and fences Limiting visitors' entrance Entrance fees		Guiding thematic tours Training special user groups Training commercial providers
Low			

As the question of "acceptance by the public" was not always investigated or known by the researchers, some countries chose not answer this question or had doubts about what answers should be given. Furthermore, "the public" consists of a wide range of opinions and is difficult to assess as a homogeneous group. Consequently, these figures are difficult to use as a general measure.

2.9 Future Trends and Challenges

2.9.1 Conclusions

This chapter compares forest based recreation in six Atlantic countries: Denmark, Belgium, Iceland, Ireland, the Netherlands and the UK. The Atlantic region, in comparison to other European regions, has a relatively low amount of forest cover and most of these forests are the result of afforestation programmes dating from the twentieth century. This means that from a cultural-historic perspective the forests and their utilisation are quite recent. Almost all of the Atlantic countries are densely populated or have a basically urbanised population. This means there is a relatively low amount of forest and other wooded land per inhabitant (FOWL/capita), resulting in great pressure on existing forests to provide the multiple benefits demanded from them.

In regard to forest policy and management, the Atlantic countries are moving away from mainly timber production to multi-resource functions and management. The increasing importance of recreation and environmental values is found in all countries. These new, additional values include environmental and social non-market benefits.

Providing public access to forested areas is an important issue. In order to increase access to forests, new areas are being planted in and around the urban centres of the six countries. The principal difference in the study areas of open public access is that in Iceland, Denmark and the UK open access is guaranteed, whereas in Belgium, the Netherlands and Ireland private property is protected. In the latter three countries, access rights are currently being redefined to increase access to privately owned lands. Public demand in Flanders and the Netherlands has also resulted in state policy changes. A movement in the same direction is found in Ireland as well.

In the six study countries, forest areas are generally small (except in Wallonia). Landowners are under increasing pressure to provide public access to these areas and to improve the recreational infrastructure. For individuals, providing these benefits may be costly and not in keeping with their personal desires. Furthermore this often entails considerable inconvenience to the landowners. Therefore this increase in public access must be accompanied by increased responsibilities of the visitors. To increase responsible behaviour by the users, several of the study countries have (or are) implementing various methods of educating the public. The methods used include such things as Codes of Best Practice, educational programmes and direct involvement with the users in the form of guided tours. The governments must also provide incentive programmes for the forest owners. Incentive programmes are found in some of the study countries, such as Flanders and the Netherlands; however these programmes must be increased to prevent the burden of provision from falling entirely on the landowners.

In most Atlantic countries (state) legislation takes into account public access of forests in relation to other forest functions. The most important forest functions in the Atlantic regions are the ecological and social functions (except in Wallonia where timber production remains an important function). These functions have somewhat conflicting goals. The fact that many natural values are found in forests explains in part the necessity of national legislation to protect these values. There is a negative correlation between the degree of legislation concerning outdoor activities and FOWL/capita in the study region. This implies the necessity of more legislation as user intensity increases.

Recreation and nature tourism in forests causes conflicts. This study has identified different types of conflicts: (i) forest recreation versus other forest functions (timber production, hunting/fishing, nature conservation); (ii) recreation activities versus other recreational activities and (iii) between recreationists. The data show that all three types of conflicts are important within the Atlantic region. Conflicts between different recreational activities are considered the most important type of conflict in all of the study countries. This is followed by the conflict between forestry use/timber use and other uses (both environmental and recreational). Finally, the lower the FOWL/capita, the greater the number of conflicts experienced.

Most conflicts are caused by activities considered intrusive, such as motorised (water) sport, mountain biking, horseback riding and orienteering. These activities can disturb flora and fauna, and other recreationists. Environmental disturbance factors, caused by visitors or forest activities include rubbish dumping, traffic noise, road construction and large clear cuts, which also affect the ability to realise social and environmental benefits from forested areas. In general, the most densely populated countries with relatively little forest, i.e. the Netherlands and Flanders, score highest for both types of conflicts. Although the UK has a low FOWL, in this country a low number of conflicts of this type are reported. This may be as a result of recreational zoning and more intense programmes of public participation in forest planning and public education. The above mentioned types of conflicts are found to a lesser extent in Denmark. These types of conflicts are fewer in Iceland and Ireland.

Another factor affecting the number and severity of conflicts is the level of recreational planning and use of planning tools such as the zoning of conflicting activities to minimise conflicts. For all countries some form of planning for recreation is undertaken, either in special cases or at regular intervals. The UK and the Netherlands report the highest level of planning. Planning in these countries is facilitated by a large amount of existing data and planning guidelines. This indicates the necessity for developing these types of planning tools to ensure that planning and monitoring are done on a regular basis. This does not however mean that planning is not done in other countries, only that it is not done in such a proactive manner. Each country was asked to evaluate the prioritisation of forest based recreation in the country's forests. Denmark and the UK recorded slightly higher than average ratings, Ireland and the Netherlands recorded satisfactory scores, while Belgium and Iceland reported lower than satisfactory ratings. This does not mean that recreation is not given high priority in these countries, but reflects that the shift from timber production to environmental and social benefits is relatively recent and the development of planning and monitoring methods are in the early phases of formation.

The data show that the recreational infrastructure in forests in urban, rural and remote areas is considered satisfactory in all countries. There is a connection between the overall level of urban character/population density on the one hand and the provision of recreational facilities on the other. In highly urbanised countries, i.e. Belgium, Denmark, and the Netherlands, the recreational provision, in quantitative terms, is considered sufficient/good. Recreational demand is so high in these countries that there is a need for a large and varied amount of recreational infrastructure. The quality is variable, but in general satisfactory. The UK is in an intermediate position. Although its population density is high, the recreational supply is only considered satisfactory in urban areas. The other two Atlantic countries, Iceland and Ireland, with a lower population density, assess the recreational supply, both in qualitative and quantitative terms, as unsatisfactory. In this respect it is interesting to observe that the Netherlands, the country with the highest population density, is the only country with a neutral opinion regarding the "over-equipment issue of forests limiting the experiencing of nature". This may be associated with increased urbanisation, and extensive provision of facilities and activities in the forests. These increases may produce a greater need for the user to experience untouched nature in forests and other natural areas.

In order to maintain and develop the recreation infrastructure and services different approaches to visitor management must be used to encourage appropriate visitor behaviour and comfort. Management actions are classified as restrictive actions (e.g. prohibitions), soft actions (e.g. infrastructure and signposting), and facilitating actions (e.g. guiding and education). Generally, throughout the Atlantic countries programmes for the guidance of visitors in forests and nature areas are implemented. Soft actions are most common and considered most desirable, followed by restrictive actions or hard actions. Overall, the frequency of management actions is highest in urban areas and decreases with increasing rurality. The differences between urban and rural areas are slight. The Netherlands and the UK frequently use a variety of actions to manage user behaviour. Denmark and Iceland use the least amount of management actions, whereas Belgium and Ireland hold an intermediate position.

2.9.2 Trends

The main findings of this study are a series of trends in the Atlantic region that affect forest-based recreation and nature tourism both in qualitative and quantitative terms. Sometimes these trends are complementary, and at others contradictory. These trends have implications for effective planning and management of forest and nature areas. The implications are interpreted in terms of pressing challenges for those involved in forest planning and management.

The following trends can be distinguished:

- *Changing demographics.* There are an increasing number of ethnic groups in countries throughout Europe. These groups may have different demands for forest recreation. Other demographic trends such as the ageing populations and changing household composition are also common within the study areas and will in turn affect the pattern of recreational use. These new and changing demands must be identified and included in recreational planning. Urban and peri-urban sprawl is leading to greater pressure on surrounding forest areas.
- *Increasing consumer expectation.* Recent economic and cultural changes in the study areas have produced a new type of recreational consumer. Recreational consumers are increasingly hedonistic and expect their personal demands to be met, despite the consequences for other users and for the forest itself. There is also an increasing demand by the users to take an active part in the planning of recreational areas, an increasing demand for access to desirable areas and for increased infrastructure within recreational areas. Although these changes are not on the whole negative, they represent new challenges for forest planners and managers.
- *Health, spiritual and environmental concerns.* In almost direct opposition to the trends of increasing demands for public access and an increasingly intensive use of forested areas, is another set of demands on recreational and forested areas. There is a demand for forested areas providing recreational benefits to provide environmental benefits as well. Other new social benefits requiring a sense of solitude within the forest include such things as the use of recreational forests

to provide sanctuaries for stress reduction to improve mental health and provide spiritual experiences. Many recreational users also turn to these areas in an attempt to return to nature and offset the negative effects produced by the increasingly man-made environments in which they live and work.
- *Safety and control*. Planners of forest recreation areas are under increasing pressure to provide for the safety not only of the forest users, but of the forest itself. These provisions and the "control" of user behaviour in recreational areas are in direct conflict with the expectations of both hedonistic users and users desiring a more natural experience within the recreational areas.
- *From government to governance*. The increases in demand and shifting emphasis placed on benefits from forest areas have made governments more dependent on the co-operation of NGOs, businesses and citizens to realise these goals. Therefore, top-down planning approaches are no longer effective. Forest management and planning must be done using co-operative participatory methods placing more responsibility on local and regional stakeholders. State, regional and local governments must also work together to produce legislation providing incentives to individual landowners and the legal framework necessary to realise these new goals.

2.9.3 Challenges for Forest Planning and Management

Complex and intense management demands governmental support. The data gathered from the six countries indicates an increasing demand for the use of forests for social and environmental benefits. These demands, together with increasing urban populations and limited amounts of forested land from which these benefits can be realised, call for more intense management systems to be developed and applied. The rise in demand also increases dependence on private landowners to provide access to their land to satisfy these demands. State involvement in reviewing and revising legislation and providing incentives for individual landowners has been done in some cases. However these efforts must be more intensive and widespread.

Multi-resource management: use of zoning to minimise conflicts. Forest management no longer deals exclusively with the profitable management of trees, but has developed into true multi-resource management. The management of forest recreation in a way that effectively addresses the desires of as many of the various user groups as possible leads to many challenges in the design and management of forest areas. Three major types of conflicts were identified: conflicts between benefit types (i.e. recreation versus environmental concerns), conflicts between users (i.e. overcrowding) and conflicts between recreational objectives (i.e. hiking versus biking). The management tool shown to be most effective for reducing these conflicts is the use of zoning of individual forests for different uses. Zoning of forests on a regional level can also be helpful. The amount of infrastructure in the forests can be adjusted between forests to allow for recreationists demanding a high level of infrastructure while in other forests or forest areas the infrastructure is kept to a minimum for those seeking more natural experiences. More intensive use of forests also calls for increased user education regarding proper behaviour in the forests.

Multi-resource management demands a multi-disciplinary approach. Forest planning that addresses the issues listed above has become a complex, multidisciplinary task. Many forest planners must deal with social, cultural, landscape and environmental issues. These are at times, outside the realm of traditional forestry education. Therefore not only must the management become more intense, it must utilise interdisciplinary expertise if the challenges of multi-resource management are to be met. In all the six countries, there is an increasing awareness of the need to include experts from fields outside forestry, for example from the social and environmental sciences.

Chapter 3
Nordic Region

Odd Inge Vistad, Joel Erkkonen, and Dan Rydberg

3.1 Introduction

Norway, Finland and Sweden are characterized by a large proportion of forest cover and a relatively low population compared with the size of each country (Fig. 3.1). There are an average of only 15 inhabitants per km^2 in Norway, 22 in Sweden and 17 in Finland. The population of each is still quite dispersed, but a centralization

Fig. 3.1 Countries within the Nordic region

O.I. Vistad (✉)
Norwegian Institute for Nature Research, Fakkelgården N-2624 Lillehammer, NORWAY
e-mail: oddinge.vistad@nina.no

U. Pröbstl et al. (eds.), *Management of Recreation and Nature Based Tourism in European Forests*, DOI 10.1007/978-3-642-03145-8_3,

tendency is dominant and probably quite similar to that in the rest of Europe. In Norway about 78% of the population live in densely populated areas, compared with about 84% in Sweden, while 67% of Finns live in towns or urban areas[1].

In all these Nordic countries[2] forest recreation (mainly traditional activities such as walking, picking berries and mushrooms, skiing, fishing, swimming, biking, hunting, etc.) is very popular and very much a part of along tradition and also part of the sense of national identity.

The general tendency for populations to become more urbanised increases the pressure on recreational resources (especially forests and coastlines) in the more densely populated areas. This highlights the need for better planning and management of multiple-use of the forests. Nature based tourism is increasing, but is only partly located in the same areas. There is not a very sharp division between non-commercial (outdoor recreation) and commercial use (nature based tourism) of forests. However, commercial activity is increasing, and it is also being stimulated by public authorities (especially in more remote areas). Nature based tourism is already important, it is increasing and it is underdeveloped. There is potential for both happy co-existence and for conflicts between recreation and tourism in addition to conflicts with other land use interests, for instance traditional timber production.

These three countries are all very important timber producers (especially Finland and Sweden). Despite the strong tradition in forest recreation, the social dimension in forest planning and management has probably developed slower in these countries than elsewhere in Europe. The situation with a lot of space, a low population and free access to any forest has probably reduced the perceived need for a comprehensive multiple-use planning of the forests as found in other countries, but times and needs are changing.

3.2 Forest Cover, Development and Ownership

3.2.1 Forest Cover

The relative forest cover in these three countries is somewhat different (Fig. 3.2), being 40% in Norway, 75% in Finland and 73% in Sweden (all from Statistics Norway 2006 – a combination of "forested land" and "other wooded land", according to ECE/FAO definitions). Finland is the most extensively forested country in Europe. Usually, there is a more strict definition of forests in these countries: only referring to forests below the coniferous timberline. This means that large areas with mountain birch (*Betula pubescens/tortuosa*) forests are often excluded from

[1]This illustrates a pattern, but the figures might not be directly comparable, according to the definition of "densely populated" and the year of measurement.

[2]By definition the Nordic countries also include Denmark, Iceland and the Faroe Islands which, according to the regional analysis regarding forest recreation fall under the Atlantic region.

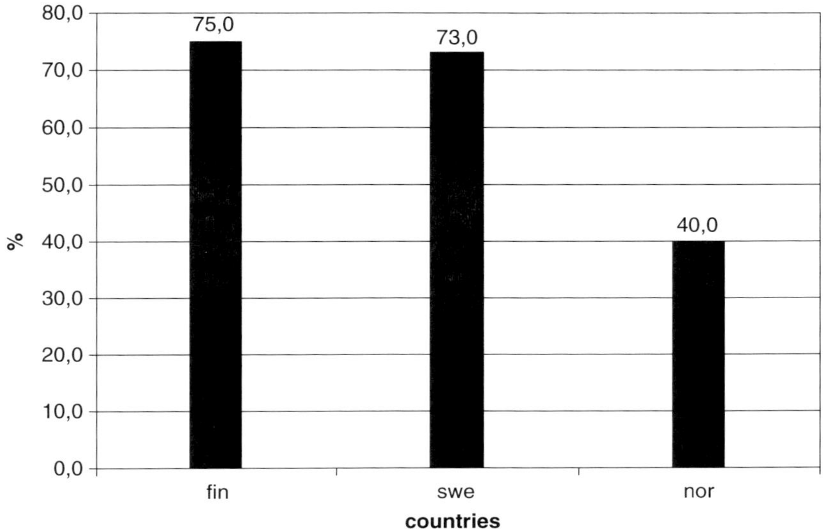

Fig. 3.2 Forest cover (%) in the Nordic countries

national statistics, but should be included so as to make the forest cover figures more comparable across Europe.

In Finland there are about 4 ha of forest for every person. Half of the forest is made up of pine (*Pinus silvestris*), the other half is mostly spruce (*Picea abies*) and birch. The same species dominate the forest of all the three countries. In Sweden about 85% of the standing forests consists of coniferous trees (about 50:50 pine and spruce). The forests are quite evenly distributed and most Swedes can reach a forest quite quickly. Only in the most southern part (Skåne) can the distance to a forest from a settlement exceed 10 km. Norway is more mountainous (44% of the land area) than the other two, and the forests are more unevenly distributed. Coastal moors are more common on the western coast and arctic landscapes in the north. However, the majority of the population live in the forested lowlands; the great majority therefore has only a short distance to walk to reach a forest.

3.2.2 Development

Climate change and reduced grazing in the mountains has stimulated the timberline (mainly birch) to increase in elevation. The general decline in agriculture has a similar effect in the lowlands – former arable or grazing lands are being reforested, either actively or more usually by natural regeneration of trees. This tendency is probably quite similar in all three countries, but the economic problems of traditional small scale farming (with consequences for reforestation) started earlier in Sweden than in Norway.

3.2.3 Ownership

More than half of the forest is privately owned in all three countries (Fig. 3.3), with a particularly high proportion – 84% – in Norway. About one third of Norway is public land, but the majority is in the mountain areas – only 12% of the forest area is public. Public forests in Finland make up 34% of the area and in Sweden 20%. Norway has 116,000 forest owners, Sweden 354,000 while Finland has close to one million forest owners. In both Norway and Finland the private properties are small (on average 60 ha and 44 ha respectively), while the private Swedish properties are much bigger (on average around 640 ha).

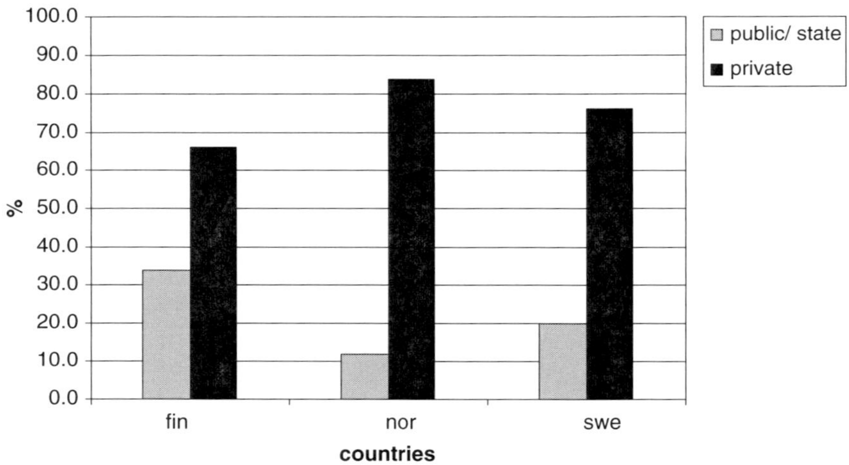

Fig. 3.3 Forest ownership in the Nordic countries

3.3 Legal Background, Public Access and Forest Functions

3.3.1 Legal Background

Only Norway has a special "Outdoor recreation Act" (dating from 1957), but all three countries have a right of free access to all forests (see next section on public access). This right is first of all a traditional, individual right (by custom), but is also partly written into various laws (Norway and Sweden), and in some cases established by courts of law through their decisions. All the countries also have forestry legislation that states the social aspects and concerns connected to forest management. In Finland this law is based on the principle of sustainability which has three equally valued elements: social, environmental and economic sustainability.

From 1994 the Swedish Forestry Act states that production goals and environmental goals are given equal importance; the environmental goals are to protect biological production, biological diversity, cultural heritage and recreational interests. The Norwegian law on forestry also states the guiding principle of sustainable management of the forest resources (value creation, securing biodiversity, caring for the landscape, outdoor recreation, and cultural values in the forest). A recent Norwegian public regulation on sustainable forestry (June 2006) emphasizes these environmental aspects and the special responsibility for the forest holder.

3.3.2 Public Access

As mentioned above: The right of free, public access to all uncultivated land (in Norway and Sweden called *"allemannsretten" or "everyman's right"*) applies in all three countries. This is an historic individual right that is only partly expressed in law. Everyman's right belongs literally to everyone – not only the locals or those who live in the country. The most explicit law is the Norwegian Outdoor Recreation Act (from 1957). This Act gives very detailed rules on what is legal and what is illegal, and the Ministry of environment has released a special publication on how the rules should be interpreted, on the landowners' and the visitors' responsibilities and so on. The visitor has to behave *"...considerately and with due care"* (in relation to the owner, other visitors and the nature). It is forbidden for the landowner to build a fence around a forest in order to keep people out, and the owner cannot require payment unless a special service is offered to the visitor. The right allows people the privilege of collecting berries, mushrooms and flowers, or to camp for one (or more) nights. The right also applies in conservation areas, unless there are special rules that regulate access. Generally the rules are quite similar in all three countries, but there might also be some different interpretations. In Norway, for instance, there are some regulations concerning where one can ride a horse or use a mountain bike. There are also differences in where and when one can make a camp fire; in Finland the landowners permission is needed, in Sweden it is legal "everywhere" if there is

Table 3.1 Existence of legislation to influence public access and other recreational uses

	Finland	Norway	Sweden
Existence of national legislation to influence public access (yes/no)	Yes	Yes	Yes
Restrictions for public access can be applied in:		**	
• Private forests	No	No	No
• Public forests	No	No	No
• State forests	No/Yes*	No	No
Existence of laws affecting recreational uses:		**	
• Sport activities	Yes	Yes	Yes
• Collection of NTFP's	No/Yes*	Yes	Yes
• Other uses	Yes	Yes	Yes

*Finland: can be applied in protected areas

** Norway: can be applied in protected areas (no matter who owns the land).

not a risk of starting a forest fire, while in Norway it is forbidden on forested land during the summer season.

The free access does not include the use of motorised vehicles, even on snow. There are special rules in all three countries, with the most difficult conditions for motorised recreation in forests and in the mountains being in Norway, where it is generally forbidden (except on special arenas, and in the northernmost part of the country).

3.3.3 Forest Functions

The main function of Nordic forests is still wood fibre and timber production, but the ecological, social and cultural heritage functions and values are being considered more and more. In some areas – especially in central and urban areas – the social values are given a high priority. "Traditional" forestry (sawn timber, paper and pulp production) is more important in Finland and Sweden than in any other European country. The production of firewood (bio-energy) is also probably increasing more than any other wood product.

It is possible that social thinking in forestry management is more "developed" or realised in Finland than in either of the other two countries – but this is somewhat difficult to judge. However, a recent Swedish study also shows that the thinking among forest holders is changing and that the social and recreational values are highlighted (Lidestav and Nordfjell 2003). The Norwegian Forest Owners' Organisation has recently presented findings showing that the present income from wood products is less than two thirds of the total forest income. The organisation has identified two additional value chains in addition to an increased use of wood, namely bio energy/bio fuels and forest based tourism.

It is quite clear that the everyman's right (free access) is a challenge in connection with the development of forest based tourism – it is difficult to make money when the main attraction (the landscape, the scenery, the entrance) is free for all.

3.4 Forest Managers' Education and Tasks Regarding Recreation in Forests

All three countries have a long and solid tradition in forestry management education – at the college and university level. However, there is probably a long way to go before the compulsory education is satisfactory when it comes to recreational and social issues in connection with forest management (Fig. 3.4). Today there are often voluntary options for developing an individual forest management education profile with quite a strong focus on recreation, tourism and experiential issues. Traditionally, social issues in connection with forestry have usually had an economic angle and are often based on activities like fishing and hunting that produce some sort of harvest.

Norway is probably quite typical in that way. When studying the forestry education program at both a bachelor and master level (at UMB – The University of

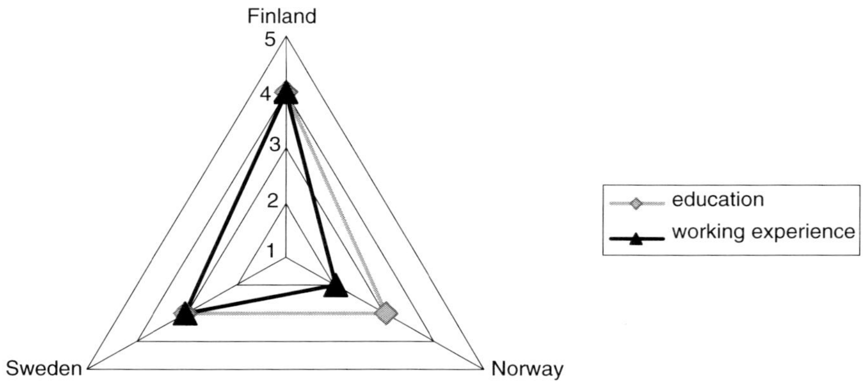

Fig. 3.4 Evaluation of knowledge about planning and management of forest based recreation obtained by education and working experience (1=unsatisfactory, 5=excellent)

Life Sciences) very little is said about recreational management or tourism development. It is just stated in the introduction that forests are the *"basis for production of wood, recreation and tourism"*. However, there is a good opportunity for students to include such topics if they have special interest, because other study programmes at UMB include recreation and tourism planning, for instance the master programme "Nature-Based Development and Innovation". A couple of regional university colleges offer bachelor programmes in forestry. Even at these colleges the special knowledge on recreation and tourism is mostly offered through other study programmes (in nature management, or nature based tourism).

At the Swedish University of Agricultural Sciences (SLU) the master degree program aims at a high degree of individual specialization, with a spectrum of courses for the students to choose from. There are courses at different levels (from basic to PhD-courses) focusing on the social values of different forest function such as recreation, aesthetics and public health. There are also different additional Master programmes in Sweden, for instance on Urban Forestry and Urban Greening, and on Nature, Health and Gardening.

The knowledge that the foresters and managers acquire through their work is very much dependent on where (geographically) they work. In areas close urban centres recreation has been a very important part of the forester's responsibility for a long time, and the need for recreational planning and management is increasing in all these countries.

Generally, outdoor recreation (hiking, skiing, picnicking, camping, berry picking etc.) in the Nordic countries is not looked upon as something that needs facilitation and management. It is an individual right – protected by law or customs, stimulated by the authorities and "without cost" for the individual – in countries with lots of nature (forests and mountains), low population and free access. Modernization and urbanization, now combined with the "crises" in traditional farming and forestry, have stimulated a consciousness of outdoor recreation and the need for planning and management. Outdoor recreation does not continue anymore only through tradition, it needs stimulation and facilitation – and therefore professional knowledge.

These days a lot of effort is put into tourism development and planning, for example by the agricultural and forest owners' organizations. Research programmes are focusing on the need for innovation and a growth in the regional tourism industry – where natural and cultural landscape qualities are the basic attractions.

3.5 Conflicts

3.5.1 Types of Conflicts

There are differences between the Nordic countries and the other regions of Europe, especially the Atlantic countries, in terms of demographic and geographic characteristics. There are also some important differences in public understanding about recreation and in its traditions, as well as other aspects of land use history. Conversely, the trends of centralization, urbanization, globalization, modernization etc. have consequences for outdoor recreation in all countries, in such a way as to suggest that recreational trends and conflict tendencies *might* be quite similar, or at least be moving in similar directions. If not, it is especially interesting to study and to try to understand why conflicts develop differently in different countries or within different contexts.

Here, five types of conflicts are systematized in the following way (Fig. 3.5): conflicts between different recreational groups (both in terms of activities and the amount of visitors), between different types of land use, between recreation and

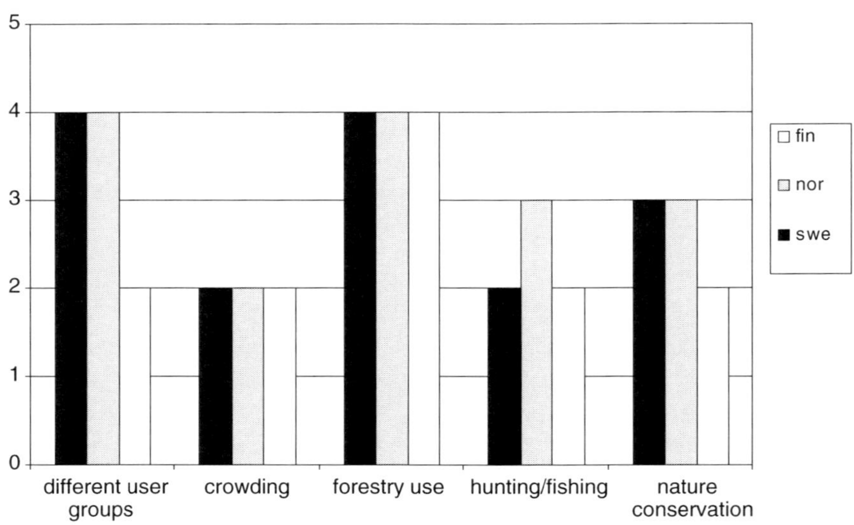

Fig. 3.5 Conflicts levels in the Nordic countries (1=not important at all, 5=very important)

other uses in the forest (such as hunting and fishing[3]), and between recreation and nature conservation/ecological interest.

Conflicts between different recreational groups or activities are traditionally not severe in these "Everyman's right countries". *Crowding* very seldom occurs. However, the development of new ("modern") and often more speedy, risky, noisy or "dramatic" activities has a tendency to develop into conflicts – often of an asymmetric type: One (often traditional) group – for example skiers or walkers – is frustrated, irritated, or annoyed over another (often modern) group – for instance snow-mobilers or mountain bikers – while the other (modern) group does not see any problem with the first (traditional) group. These types of conflicts will probably increase. One reason for this is that more and more recreational activities seem to take place in the same area. This is an effect of centralisation and a fast rate of development of new activities within recreation. The problem is also worsened by a lack of tradition in handling social conflicts, and even a lack of good management tools for handling or regulating them. It is easier to "attack this problem" when an activity causes damage to the natural environment (damage to the ground, disturbance to animals etc.), than when it causes negative social impacts. Under the typical Nordic forest-recreation-everymans'-right conditions recreational management is not believed my many to be necessary, and (social) conflict management has not really existed as a practice. This might be a more position for some of the countries (typically Norway?), but on a general level this is a characteristic of "the traditional Nordic recreational management culture".

Hunting and fishing is very much a part of the "recreational family" of activities. Hunting might be controversial in some urban areas, but it is seldom a burning issue. Hunting (and fishing) has a high social profile in the Nordic countries. Traditionally, both hunting and fishing are connected to the land owners' rights, and is a part of the traditional farming and subsistence/survival economy in rural areas. However, a social profile developed, especially through the twentieth century; hunting and fishing are not accepted as reserved rights for land owners or rich people (although it is not a part of the everyman's right, like other types of harvesting, for instance picking berries or mushrooms). Hunters' and fishermen's organizations and other interest groups protect opportunities for ("cheap") hunting and fishing for those interested. This is very much a political issue, not only a question of market concern, and is generally well supported by the public.

Conflicts between recreation and nature conservation are of principal concern in all the Nordic countries, though the problem is seldom highlighted in recreational areas/"normal forests". It is first of all an issue in protected areas with potentially threatened natural qualities. However, in some forests the problem might be reversed: conservationists might prefer to allow natural ecological development and disturbance processes in the forest (e.g. leaving dead trees overhanging a trail), while recreational users might prefer more open, tidy and managed trails.

Clean, open and (even) reconstructed trails are also an important issue in the *relationship between forestry activities and recreation*, because after logging has taken

[3]Fishing and hunting are usually categorized as recreational activities in the Nordic countries.

place *trails* can be covered in branches and severely damaged by logging equipment. It is usually necessary to clean up after the logging is completed. The forest holder is not always the logger or the forest worker. For example, in Norway it is the forest holder who has to make sure that foresters pay attention to environmental concerns, including cleaning up in and along *"often used paths, trails or other traffic arteries"* when the logging is finished. Any damage after the logging should also be repaired as soon as possible. However, this trail issue is probably more a potential conflict than a common actual conflict between forestry and recreation. In some areas, especially in urban forests, the use of *clear cutting* is more of an actual conflict – primarily for aesthetic reasons. Forestry efficiency and forester's preferences might meet resistance from recreational interests. Building *forest roads* is another controversy – on the one hand it improves accessibility for walkers, but on the other hand, when the road density is "high enough" for access reasons, continued road building ("necessary" for forestry) might negatively influence the recreational experience. However, in Norway the number of new roads (in km) constructed for forestry reasons has decreased dramatically over the last fifteen years – in part replaced by reconstruction of existing roads. In addition the total amount of forest roads is very high. Therefore, has this development occurred because the general need for forestry roads has reached saturation point?

3.5.2 Important Activities and Their Conflicts with Nature Interests

When different recreational activities and their level of conflict with nature interests are examined, the overall picture is a "low conflict level". There are of course some exceptions (with medium conflict level, see Fig. 3.6). There are different opportunities to undertake *motorized recreation* in the three countries, and also some regional differences. Both the actual and potential consequences for natural qualities are quite significant. However, the type of motor vehicles together with situational, seasonal and other natural differences influences the effects. For vegetation it is of course travelling on bare ground that has the greatest negative impact, especially over bogs and on wet or extremely dry ground. Such activity is mostly restricted, but it is for instance relevant in Northern Norway. Good planning and well prepared tracks can limit the damage. Snowmobiles have less effect on the ground, but there are some potential threats when the snow cover is thin. However, snowmobiling is a much more common activity than the use of (bare ground) all terrain vehicles (ATVs) – especially in Finland, Sweden and Northern Norway. In southern Norway it is forbidden (for recreational purposes). The effects on wildlife are far more complicated than they seem at first sight and it is difficult (and often wrong) to generalise from an impact on one animal species to another. The immediate interaction between vehicle and animal dominates the research on effects, but the most important are the long term, cumulative effects (together with other kinds of disturbances) on population or ecosystem. Calculating such effects is very difficult to design and implement. Many researchers are warning managers and recreationists

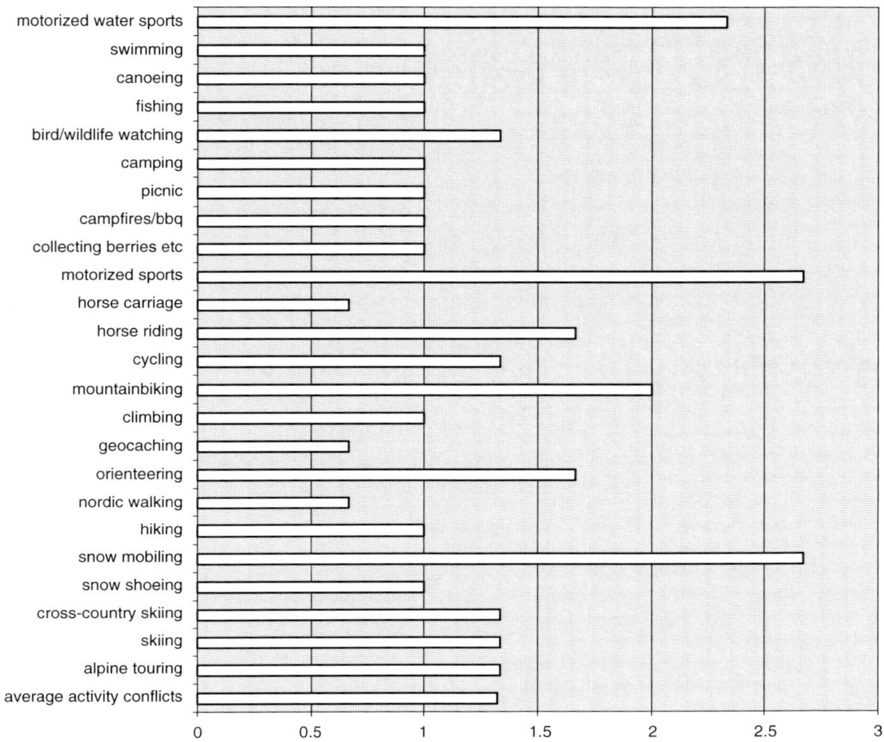

Fig. 3.6 Frequency of conflicts with recreational activities (0=non-existent, 1=rarely, 2=medium, 3=frequently)

about this, especially because motorised vehicles have a very wide geographical range (compared to other recreational activities). Good planning of motor tracks can of course limit the negative effects. Such effects may be under a particular research focus in Norway – because of the unique Norwegian responsibility for the wild reindeer – but this focus is more relevant in the mountains than in the forests.

Mountain biking is the one modern (bare ground) activity that really has increased in popularity and size. The ecological effects can be widespread – especially on the ground and on vegetation, often intensified by erosion from running water. There might be different opportunities to regulate mountain biking in the three countries. The establishment of special down hill arenas for mountain biking can limit both social and ecological problems – when well planned. *Horse riding* can cause similar problems, but horse riding is more limited in popularity.

Orienteering can also cause some ecological problems, especially during big competitions. However, good planning and preparation can reduce the effects on both fauna and vegetation. In Finland, especially, *climbing* is reported as an activity that causes some cliff erosion problems. In Norway making *campfires* is particularly mentioned as a risky activity, because it causes forest fires. It is forbidden in the summer season but is still a part of the recreational culture and many people do not respect this regulation.

3.5.3 Impacts and Disturbance of Other Societal Activities on Recreation

All three countries report negative effects from other societal activities on forest recreation (Fig. 3.7). Some of them are already mentioned (connected with forestry activities).

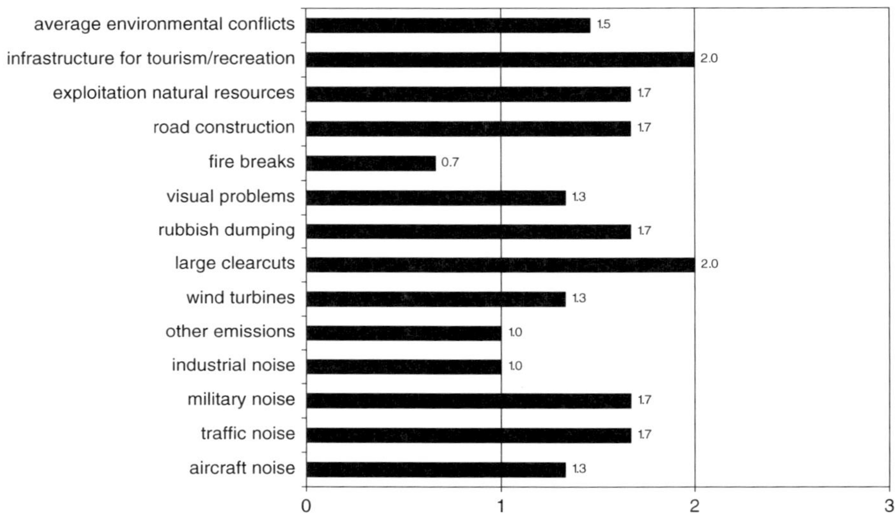

Fig. 3.7 Frequency of environmental conflicts in the Nordic countries (0=non-existent, 1=rarely, 2=medium, 3=frequently)

One of the main problems – reported as medium significance everywhere – is *noise* caused by different activities or sources, such as *aircraft, traffic and military activity.* However, the severity is generally rated higher than the frequency of the problem. One exception is traffic noise which is reported as quite frequent in Norway and Sweden, but not severe in relation to forest recreation. Industrial noise is not reckoned to be a problem. Sweden also reports *other emissions* (odour, dust, pollution) as of medium importance but not very frequent. Both Norway and Sweden report *rubbish dumping* (old cars etc.) as both medium-severe and frequent.

The two other problems of medium importance are connected to *exploitation of other natural resources*, and established *infrastructure for tourism activities* (like ski lifts, cable cars etc.). For both, the severity is greater than the frequency, except for tourism infrastructure in Finland which is also medium in frequency. From Sweden also *visual problems associated with damaged forests* (meaning severe storm, insect or fungi damage) are mentioned as medium in both seriousness and frequency.

3.6 Planning, Management and Monitoring

3.6.1 Planning for Recreation and Nature Tourism

In the Nordic countries it is probably important to remember the following, as an introduction to this section: because of *"Allemannsretten"* (Every man's right – free public access) and because hiking is the dominant activity – in forests with plenty of space for everyone – the recreational opportunities have generally always been good. Conflict levels have also been low, partly because "there is always somewhere else to go hiking". The (perceived) need for recreational planning and management has traditionally been low, even though outdoor recreation is very much a part of the Nordic culture. However, *"Allemannsretten"* makes it difficult to handle social recreational conflicts when they do appear. Good recreation planning and regulating tools have been missing from manager's portfolios. Social conflicts are increasing, because of centralization, increased exploitation of forests in popular regions (close to the cities), more recreationists and more recreational activities "competing" or co-appearing in the same areas and so on.

Table 3.2 and 3.3 give an overview on planning for recreation and nature based tourism in the Nordic countries.

Since 2003 all "urbanised" Norwegian municipalities are encouraged to develop urban forest plans. However, no municipality is obliged to develop such plans. This planning is usually based on a planning need developed through great public recreational pressure on rather limited forest resources, for example around cities. In these areas traditional forestry has – for decades – been balanced with the experiential needs of the public. This planning has been based on the traditional free access and is not meant for commercial development – except for special rules concerning hunting and fishing, accommodation etc. Trail construction (including ski tracks) and maintenance is very much a part of planning. In the recent (2006) Norwegian guidance for sustainable forestry significant responsibility is laid on the individual

Table 3.2 Availability of independent planning (legally required or optional) in the Nordic countries

	Finland	Norway	Sweden
Independent planning for recreation and nature tourism in forests	Yes, at regular intervals	Yes, for recreation in urban forests	Yes, in special cases
Legal or optional requirement for forest owners	Optional (legal req. for PAs)	Optional	Optional

Table 3.3 Steps that are undertaken for planning forest recreation at regular intervals in the Nordic countries

Steps in planning forest recreation	Countries*
Inventory of ecological conditions	Fin, Nor, Swe
Inventory of social conditions	Fin, Nor, Swe
Analysis of conflicts	Fin, Nor, Swe
Definition of vision and goals	Fin, Nor, Swe
Participation in the definition of goals	Fin, Nor, Swe
Definition of standards	Fin
Deduction of measures	Fin
Participation in the deduction of measures	Fin
Monitoring	Fin

* Norway: only relevant in the optional recreation planning in urban forests.

forest landholder. However, this is not a forestry planning instruction; it is a set of instructions concerning environmental issues (including recreation interests) that the forest holder should keep in mind when he or she is implementing forestry actions.

There are some handbooks on planning, management and facilitations for recreational purposes in forests, for example DN (2006, 2003). Both The Planning and Building Act, The Outdoor Recreation Act as well as the forestry acts are relevant for such planning. According to the forest acts forest owners/forest workers are generally obliged to take "due care" of public interests when cutting timber etc. According to the law, existing paths/trails/roads should be kept open for the public, but it is assumed that this very much depends on the will and level of "public consciousness" among the forest owners. In Urban Forest Plans there are special guidelines concerning practical forestry and timber cutting, in order to minimize negative impacts on outdoor recreation and aesthetics.

The Forestry Act in Sweden (dating from 1993) is a minimum requirement act where the private forest holder is expected to provide further voluntary initiatives within the spirit of the act. For the forestry sector, freedom, responsibility and knowledge became key words. The forest land-owners obtained increased freedom to make their own decisions about their forest, but with the responsibility to reach the broader forest political goals. The aesthetic and social values should be protected according to the environmental goal. In Sweden, multiple-use planning is widely used in forestry. In the multiple-use plan the different uses are regulated to reach an optimal solution for all uses. However in traditional multiple-use planning, individual recreational needs are not specifically described. There is still a lack of useful planning methods for forest recreation areas in Sweden. Despite the fact that there are new forest management plans adapted for recreation available for some years, there is still a need for development. This requires a shift from the traditional focus on forest stands, stand age and species composition, towards one where the total experience and system of paths is considered. The planning should also include the whole nature area not only the forests.

There are quite a few recently published handbooks about recreational planning and management in Sweden. In 2005 the County of Stockholm published a handbook discussing recreational planning in urban nature (Ahlström 2005). The same year the Swedish Forest Agency produced handbooks concerning the management of urban nature, access to the forest for disabled people and visitor studies in nature areas (Andersson and Rydberg 2005, Lindhagen and Ahlström 2005, Lundell 2005, Rydberg and Aronsson 2004). In 2006 the report Best Practice was published which gives advice and help when building in nature areas for visitors (Rosengren and Franzén 2006).

Forest policy in Finland is translated into practical action by various means, including forest planning and the activity of forest organisations. The target areas of forest planning range from single farms to regions and the whole country. The highest forest authority in Finland is the Ministry of Agriculture and Forestry. The duties of regional forest authorities are managed by the 13 Forestry Centres. Among other things, they enforce the Forest Act and also manage the forest planning for regions and individual farms. Forest Management Associations are statutory organisations of private forest owners. They operate within one or several municipalities, and their aim is to support the profitability of forestry and the achievement of other forest management goals set by the owners. Forest owners are subject to a statutory forest management fee, which is used to finance the advisory activity of the Forest Management Associations. The associations are entitled to levy a payment for other services which they provide for the forest owners.

The authority in charge of managing the state forests of Finland is the Metsähallitus. Certain sections of the state forests are in commercial use, while others are protected. Under the Metsähallitus organisation, the management of these two is strictly separated. The management goals of commercial forests are defined by the Ministry of Agriculture and Forestry, and those of protected forests by the Ministry for the Environment. The aim of commercial forest management is to generate revenues for the state, while the management of protected areas is financed by the state from its annual budget. Metsähallitus uses a multi-stage planning system covering regional decisions on land use as well as detailed local plans for a particular operation. Natural resource planning means that the use and management of natural resources is broadly viewed from the perspective of their different forms of use. Different forms of use of state land are nature conservation, forestry, recreation, nature tourism, real estate development and the sale of soil resources. Land-use and management plans are devised for the nature conservation, wilderness and hiking areas that are under the administration of Metsähallitus. With the help of these plans, the goals of nature conservation, recreation and other uses for areas are reconciled. The decisions made in natural resources and management planning are implemented at stand or site level by means of detailed operational planning. Operational planning in forestry includes silviculture and felling plans, while in protected areas, planning mostly includes habitat management, forest and mire restoration and route plans.

Today there is more and more interest in alternative ways of using forest resources, especially in the direction of commercial recreational (particularly concerning fishing and hunting). There are also pilot studies and reports exploring the potential, need, possibilities and encouragement of tourism development in forests and mountains. The Norwegian Forest Owners' Association has been active in many of these processes, and also other forest and land owner organisations, often in cooperation with tourist associations. The main effort here is not primarily to make good plans for the public, but to try to make money from tourism based activities in the forest. Even the Every man's right has been questioned in this connection. But the Managing Director of the Norwegian Forest Owners' Association has removed all doubt, and said (in 2005) a definite "no" to charging for forest access; forest owners have always guaranteed free access to nature for the public.

In Sweden it is a similar situation. During recent years the importance of regional and local development has been stressed. Every county has a regional agreement designed to contribute to the improved collaboration between regional and local bodies. The goal is to make the most of the particular conditions in each region and thus promote the type of long-term growth that favours the creation of new companies and new jobs. Since Sweden is a land of forest, opportunities to develop new forest enterprises are of great interest. The potential to develop the sector of nature based tourism has been stressed. As a part of the urban nature's potential for regional development, the Regional County Boards have an obligation to identify and protect urban nature areas of recreational importance. Sweden is currently the only country in the northern hemisphere having a segment of institutionalised, i.e. certified ecotourism. Nature based tourism entrepreneurs, such as Sveaskog (the state forestry enterprise), could be members of the Swedish Ecotourism Association under the quality label "Nature's Best". The main organisation backing the quality label is Swedish Ecotourism Society and, together with the Swedish Travel and Tourism Council and the Swedish Society for Nature Conservation, their common objective is to see ecotourism in Sweden grow bigger and better.

In Finland the National Outdoor Recreation Demand and Supply Assessment (LVVI) was carried out in 1997–2000. The aim of the study was to assess and to gather information on outdoor recreation demand and supply nationwide in Finland. A large population survey (sample size 12,000) was the principal method of data collection in the demand study. In a supplemental study, a questionnaire of recreational infrastructure was sent to every municipality in Finland as well as other organisations related to outdoor recreation. The LVVI study also produced a standardised method for visitor surveys which has been widely used, especially in Metsähallitus. Statistics Sweden has, since 1975, continuously performed a Living Conditions Survey (ULF), and similar national surveys are regularly implemented in Norway. These surveys gather, among other topics, information on outdoor recreation demand and use (Statistics Sweden 2004, Statistics Norway 2004). In 2005 the Swedish Forest Agency ordered a survey on the forest actors' preferences, attitudes and visions towards the Swedish forest. The social value of the forest was one important component in the study (Kairos Future 2005).

Otherwise, most of the inventories/studies in recreational areas are done as a component in university or college education, or in research projects, both in Sweden and Norway. But social monitoring in recreation areas is about to develop in these countries. Finland has an established monitoring system (Metsähallitus). Sweden is also developing and improving different methods and standards for more systematic social monitoring (Naturvårdsverket); it started with social inventories in the green structure in urban areas. The information is then used for planning in these areas. Norway has a longer way to go concerning social monitoring, but a couple of counties have established monitoring programmes in some "hot spots". Monitoring programmes should be long term and systematically implemented, and therefore the word "inventories" is often more suitable for what is going on. The Norwegian Nature Inspectorate (SNO, established in 1996) has ambitions in the direction of social and biological monitoring – but firstly aimed for nature protection areas (and mostly in the mountains).

During 2004–2007 a group of experts from Nordic and Baltic countries produced a manual called *Visitor Monitoring in Nature Areas*. The manual is an outcome of a joint effort of experiences among participating countries, and other existing material. The aim of that work is to establish a standard approach to visitor monitoring in nature areas across Nordic and Baltic countries (Kajala et al. 2007).

3.6.2 Overall Evaluation of Recreation and Tourism in Planning and Monitoring

Each country was asked to evaluate the integration of forest based recreation and nature tourism in the planning and monitoring of the country's forests. From the overall evaluation of forest-based recreation and nature tourism in the forests of the Nordic countries it can be seen that the current state was evaluated as improving, but still not satisfactory. Also, public efforts have been more concerned with recreational interests than with forest-based tourism. Both recreational planning and monitoring in forests seem to be the best developed in Finland, while Norway, especially, has a long way to go – with the exception of recreational planning in the most important urban forests. A good Norwegian example is this: In 2008 a national act for Oslo Urban Forest ("Oslomarka") will probably be put forward, in order to designate Oslomarka as a conservation area for outdoor recreation reasons – a unique proposal in Norwegian conservation history.

3.7 Recreational Infrastructure

The basic infrastructure for recreation in forests is the trail or the ski track, combined with access roads and parking facilities. The quantity or length of trails (in km) is quite satisfying in all three countries, and in most types of areas – from the urban forests to the most remote mountains. However, the quality of the trails may vary considerably. Some are well prepared (mostly in more urban forests) and suitable for a broad spectrum of visitors (including disabled people), others are more primitive. The reason for this satisfying situation (Figs. 3.8 and 3.9) is not necessarily because a lot of work has been put into trail construction and maintenance. The Nordic countries have a long and continuous tradition in outdoor recreation, especially hiking. In fact the trails are often themselves a result of walking/hiking,

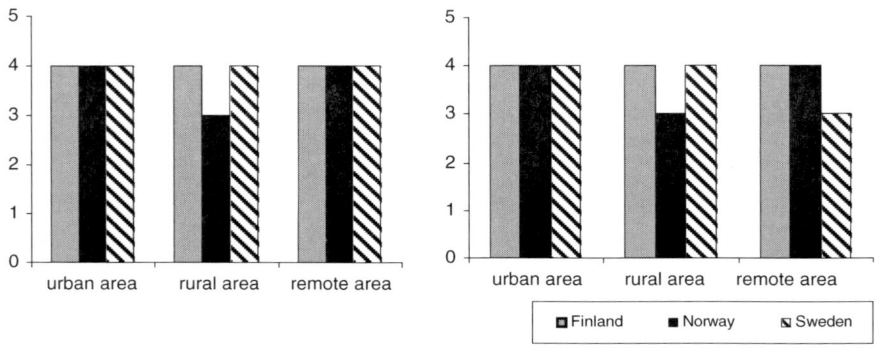

Fig. 3.8 Supply of basic infrastructure (*left figure*, **a**) and additional infrastructure (*right figure*, **b**) (no bar = non-existent, 1=unsatisfactory, 5=very good)

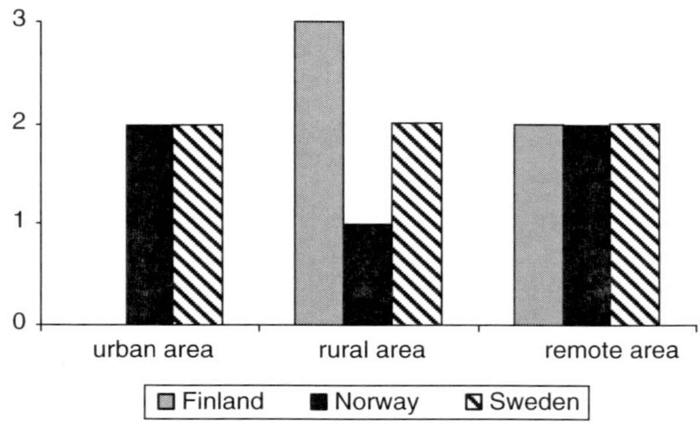

Fig. 3.9 Quality of infrastructure for recreation and nature-based tourism (no bar = non-existent, 1=unsatisfactory, 3=excellent)

and not a premise for use. This is especially the case in remote and many rural areas. During recent decades a lot of existing trails have been waymarked, primarily to help and stimulate new user groups, and to help them feel safe. In other cases the trails are well prepared and marked in order to avoid the visitors walk off the trail (especially for conservation purposes). Often there are also cabins/cafes in the urban forests where (usually simple) meals and hot drinks are served. These cabins are often the goals for family walks at weekends. Benches and tables are often placed along these trails, and sometimes even toilets are provided.

The present situation concerning existing recreational infrastructure (especially trails) is most satisfactory in the urban areas/forests and in the most remote areas (usually the mountains) – at least in Norway. The tradition with long distant hikes is typical for all three countries. The Norwegian Mountain Touring Association (DNT) is one of the oldest of its kind, in the world, established in 1868. It runs more than 400 cabins (staffed and non-staffed) with sleeping accommodation. Some of them are also located in forest areas, and quite often these cabins are on privately owned farms or mountain farms with a formal agreement with DNT. In addition

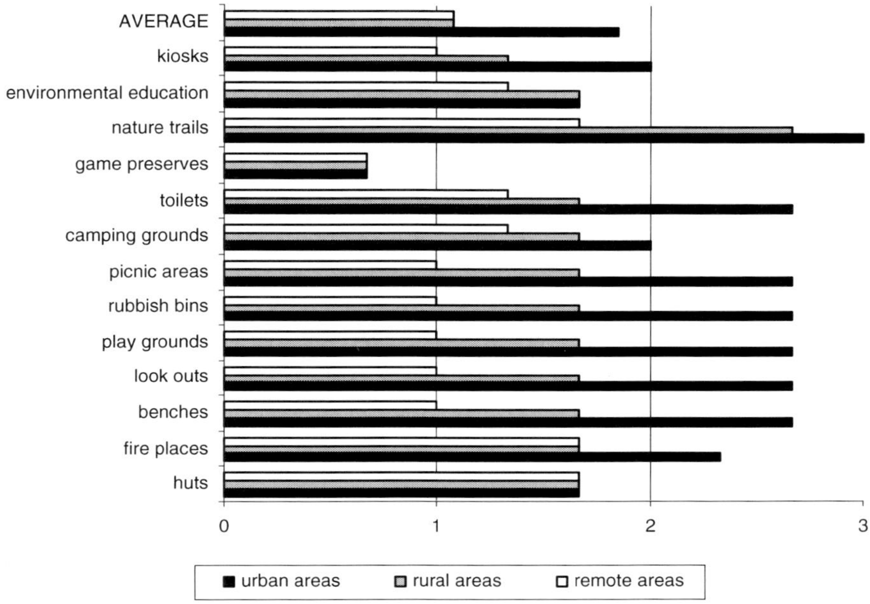

Fig. 3.10 Occurrence of additional infrastructure in urban, rural and remote areas in the Nordic countries (0=non-existent, 1= rarely, 2=reasonable, 3=frequently)

DNT has the responsibility to mark and maintain 20,000 km of summer trails and around 7,000 km of ski tracks (quite simply marked with (usually) birch branches). In Sweden there are more than 100 different hiking trials crossing the country, with a total length of more then 13,300 km. Most of the hiking trails cross privately owned land where the independent landowners have generously agreed to let the trials pass though their land without demanding any compensation. Normally, agreements are reached between local councils and landowners who, in some cases, also give their consent for rest stops or overnight facilities. In Finland recreational areas and hiking trails as well as protected areas create the most important recreational services for the citizens. Recreational areas are usually divided into urban areas, recreational areas and hiking areas. In addition, there are hiking trails that are located both inside and outside the recreational areas. Municipalities manage about 80% of Finland's recreational areas and every municipality in Finland has about 10–14 areas reserved for recreation. These recreational areas are usually rather small in size, but they have a significant influence – with marked trails and free facilities – on the quality of environment and the standard of Finns' everyday life. Some of the best areas for hiking lie within national parks and hiking areas.

It is characteristic for these countries that almost all outdoor recreational services and facilities are free of charge to the public. From Finland the following recreational facilities are especially significant and popular:

- nature trails with information presented on notice-boards or in free booklets
- picnic sites with tables, cooking shelters and campfire sites with free firewood
- free facilities such as wells and campfire sites in areas reserved for free camping
- shelters where visitors may sleep overnight
- steps and boardwalks along popular trails to prevent erosion
- compost toilets
- waste collection points

The type and standard of facility might differ from country to country, from area to area, and from trail to trail. And some facilities are more controversial than others, for instance campfire sites, waste collection/litter boxes, and toilets. It is somewhat difficult to rate the quality of infrastructure since type and amount of infrastructure is sometimes controversial – even among the recreationists. A highly facilitated recreational area (with good technical facility quality) is not necessarily regarded a high quality recreational area! Some recreationists say that facilitation disturbs their nature experience. Figures 3.9 and 3.10 must be interpreted with this in mind.

3.8 Approaches to Visitor Management

3.8.1 Restrictive Management Actions

Once again, the "Every man's right" very much limits the use of restrictive actions, but there still are some examples of such actions applied in Nordic countries. Generally, entrance fees are (at present) irrelevant as tools in all three countries. A serious discussion about introducing them would probably raise a storm of protest. In addition, powerful recreational organisations defend this right as the most sacred element in Nordic outdoor recreation. However, there are of course situations when restrictions are necessary, and probably even accepted, especially in connection with strong (and well understood) conservation interests (Fig. 3.11). There seem to be some differences between the three countries. From both Sweden and Norway it is reported that such regulations are rare – from urban to remote areas – and there is also a generally low acceptance for such actions. In a Norwegian study (Vistad 2003) among managers and visitors, looking at management of recreational impact on the ground, there was no standard answer concerning the use of prohibitions and regulations. Sometimes such actions were accepted, other times not: Specific regulations (concerning certain types of behaviour in specific areas) were more accepted than general regulations (for instance to prohibit large groups of visitors to wildland areas).

In Finland there seems to be more use of some restrictive actions (rules, regulations, prohibitions and taboo zones) in remote areas – and the restrictions have a medium level of acceptance; taboo zones are even highly accepted. Regulating or forbidding access is probably accepted in all the countries, when the reasons behind

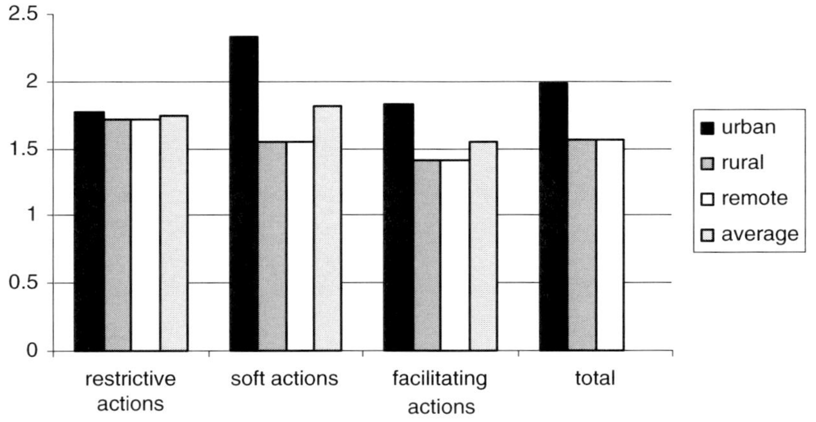

Fig. 3.11 Occurrence of management actions in the Nordic region, in general and in urban, rural and remote areas (0=non-existent, 1= rarely, 2=regularly, 3=frequently)

such actions are well understood. In Norway the impression is that some restrictive actions might be more acceptable in urban areas than in rural or remote areas. The reason for this is that the urban population is more used to active planning, facilitation, and regulation concerning outdoor recreation (Fig. 3.11 and Table 3.4).

This anecdote from Norway might illustrate some important aspects about recreational management also in other Nordic countries: If there is a Norwegian visitor "management tradition", then it is to do nothing. And if we continue to speak in general terms, and slightly disrespectfully perhaps, we could say that there has been low effort, low funding and a low consciousness concerning outdoor recreation planning and management. This is the reverse side of having a deep outdoor recreation history

Table 3.4 The level of acceptance in the Nordic region for the various management actions (high >= 2.5, medium > 1.5 and < 2.5, low =<1.5)

Acceptance by the public	Restrictive management actions	Soft management actions	Facilitating management actions
High		Attractive infrastructure Educational infrastructure	Guiding thematic tours Training special user groups Training commercial providers
Medium	Rules and regulations Prohibitions Marking protected areas/tabuzone	Signposting responsible visitor behaviour	Guiding ranger systems
Low	Entrance fee Barriers and fences Limiting visitors' entrance		

and culture, with strong and unquestioned individual rights concerning both access and behaviour in forests and mountains. However, if a problem or a threat is given political priority, some say that the typical Norwegian reaction is to introduce a prohibition, or to make strict rules and regulations. Sceptics will say that this is because it is the cheapest way of handling a problem, since the rules are seldom followed by funding, or a staff to manage or control the situation. Two typical examples are the prohibitions of campfires during the summer season in forested areas (a rule that is not really respected – unless in dry periods), and the prohibitions of snowmobiling and other motorised recreation in nature areas (a rule both loved and hated). But Norway is slowly moving out of this low-effort and low-conscious situation.

3.8.2 Soft Management Actions

Such actions (improvement of the attractiveness of areas, education of visitors etc.) are used quite often in the Nordic region – especially in Finland (in both urban, rural and in remote areas). The acceptance of such actions is also generally high. Sweden shows a similar situation, but such actions are more typical of urban areas. In Norway the situation is probably a little bit different (less frequently used, and possibly less accepted), but both the use and acceptability is higher in urban areas than elsewhere. The study mentioned earlier (Vistad 2003) showed that information and education can be very well accepted in Norway. However, the use of facilities is probably somewhat more controversial in Norway compared to both Sweden and Finland, especially in rural and remote areas (Vistad 1995). It still seems to be easier to accept facilities when the reasons are ecological (to protect the environment) and not "only" to support recreation interests – for instance the use of board walks.

3.8.3 Facilitating Management Actions

The recreational areas (forests and mountains) in the Nordic countries are not well staffed – often quite the opposite (at least in Norway). There is probably a general tendency to develop the (sparse) ranger or warden system from primarily a control function to an educational or guiding function. It is probably more the (private) commercial interests (tourism) that stimulate guiding activities. These functions can be quite diverse: nature experience, training different visitors or user groups, or ecological learning. Such activities are probably quite acceptable, but the expressed need (among local visitors) is still quite modest. Guiding and education will probably be strongly stimulated when nature based tourism develops – because different visitor segments have different needs (Table 3.4).

3.9 Future Trends and Challenges

There is no doubt about the developing importance of social values in forest management. These social aspects will also probably become more and more diversified and specialised. All three countries particularly noted the increasing interest in developing nature based tourism. This is partly connected to the need for new employment because of economic problems in many rural areas, but also connected to the developing position of leisure and recreation in (post-) modern life. There are big challenges in fulfilling the spectrum of (social) needs and also combining these social aspects with more traditional forestry and the new growth in bio-energy production. In addition to tourism, bio-energy is probably the other main "new" product seen as providing an optimistic future for forestry. This optimism is also a challenge, because it is a sector with new conflict potentials (concerning social values).

In Finland the VILMAT Action Plan aims at promoting the development of nature tourism and outdoor recreation to provide double the number of jobs in the field by 2010. The VILMAT Action Plan has a significant influence – among several ministries – on Metsähallitus, the Forest Research Institute, Finland's environmental administration and The Finnish Tourist Board (MEK).

Therefore, Metsähallitus has drawn up plans to develop the recreational and nature tourism services in the state-administered protected areas, public waters and recreational forests. Principles for sustainable nature tourism have been formulated and are used as tool in co operation with entrepreneurs. Criteria and indicators to measure sustainable use are being developed. Actions aim to increase the number of visits to recreational and protected areas without risking nature or culture values.

The Norwegian Forest Owners' Federation has led a study of the present income structure of Norwegian forest holdings and an analysis to identify the main income potential. Surprisingly, in a wood production-focused forestry management they found that the current income from wood was less than two thirds of the total forest income. They also found that the highest potential for future income increase for the forest properties was in the further development of three value chains: increased use of wood and the wood-working sector, bio-energy including bio-fuels, and forest based tourism. The situation in Norway is characterised by a combination of optimism and scepticism towards this new social (tourism) focus in forest management. And there is probably a longer way to go towards a successful forest based tourism (except from the well known products within fishing and hunting) in Norway, compared, for instance, to Finland. Part of this forest-based tourism will have to be based on more infrastructure, a stronger facilitation and more active stimulation than has characterised Norwegian forest recreation to date. This means that there will be no flying start, and new possible conflicts. Facilities for recreation (and for nature tourism) are very much lacking and are often also consciously avoided, because they are not looked upon as being necessary in the Norwegian (rural) recreational tradition. There is also a lack of knowledge on recreational use of the mountains and the forests – there has been a lack of funding for user studies, a lack of tradition of social monitoring, and it is also more complicated to monitor use when free access applies. It will be difficult to judge and decide on carrying capacities. There is also a lack of knowledge on the potential (commercial) market for new nature and culture based tourism products. Research is needed in many areas.

In March 2005, the Swedish Forest Agency formally adopted a set of objectives for the nation's forest sector. They include overall policy objectives laid down by the Swedish parliament, plus thirteen quantitative targets. The process of developing forest-sector objectives with a strong emphasis on stakeholder involvement satisfies many of the requirements of national forest programmes. Three of the quantitative targets concern multiple use and social values. One of the targets is recreational management of urban forests with an agreement between municipalities and the Swedish forest agency about the entire community's long-term ambitions regarding utilization of urban forests to improve levels of satisfaction with the supply of recreational areas. There is also a national programme of education regarding special considerations for the social values of urban forests. The main challenge for the future is how best to introduce outdoor recreation and nature as an important component for the regional and local growth and how to find a sound balance between the different uses of forests, the balance between exploitation and protection/conservation (which has increasing public support), and also find a proper planning and management approach adapted for recreational needs.

Chapter 4
Central Region

Peter Breman, Bruno Baur, Renate Bürger-Arndt, Tessa Hegetschweiler, Marcel Hunziker, Olivier Picard, Ulrike Pröbstl, and Veronika Wirth

4.1 Introduction

Central European countries described in this report are Austria, France, Germany and Switzerland (Fig. 4.1). Other regions like North Italy would also be a part to this group but the whole country shows mainly characteristics of the Mediterranean region. The Northern part of Germany and France also show features of the Atlantic

Fig. 4.1 Countries within the Central region

P. Breman (✉)
Office national des forêts, Direction de l'Environment et du Développement Durable, Boulevard de Constance, F-77300 FONTAINEBLEAU, France

U. Pröbstl et al. (eds.), *Management of Recreation and Nature Based Tourism in European Forests*, DOI 10.1007/978-3-642-03145-8_4,

region and the Southern part of France shows characteristics of the Mediterranean region. However, these countries are mainly characterised as Central countries and, owing to a lack of differentiated data it is not possible to subdivide them

The Central countries are generally rather densely populated. There are both densely populated agglomerations and scarcely populated rural areas; these regional differences are expected to increase further over time. In Germany there has been an east- west- shift of the population since re-unification that is still continuing.

In Central Europe generally the use of forests for recreation is of great importance. The very emotional discussion of the "Waldsterben" (forest dieback) in Germany underlines this. Studies have shown that if people are asked to describe an optimal landscape for hiking and cycling, forests are always included (e.g. Wöbse 2002).

Forests also have a great importance for the tourist offer in the Central countries. The well known tourism destinations in Central Europe (except city-tourism and coastal tourism) are mainly located in areas with a high proportion of forest such as the Alps, the Black Forest, the Vosges mountains and many others. For these areas, which were once economically underprivileged regions, tourism has provided the chance for economic improvement. Due to the combination of nature and landscape beauty they are now attractive tourism destinations. Today, forest and landscape is the base for different offers especially for elderly people, families, sports-orientated people and for nature-based tourism. Tourism plays an important role in these regions. In the past, forestry has often been a partner in the development of recreational infrastructure such as winter sport facilities, trails, huts and information centres.

The mostly mountainous forests within the Central region are also characterized by a regional culture that shows their close relationship to forestry and life in or close to forests. Most of these regions are known for special wooden handicrafts such as the cuckoo clock from the Black Forest, the Christmas decorations and candles from the Erzgebirge or wooden instruments from Switzerland. Architecture, especially old farmhouses are characterized by various traditional forms of wood construction, wooden facades, roofs, windows, balconies and so on. The use of wood in construction and architecture still plays an important role in most of these regions. Private forest owners are therefore unwilling to give up their land even they do not have enough time to use their forest in an adequate way.

4.2 Forest Cover, Development and Ownership

4.2.1 Forest Cover

The average forest cover in the Central countries is about 35% of the total land area, similar to the average for Europe. Austria has the highest percentage of forest cover at 47.2%. Forest cover in Germany is 33.0%, in Switzerland 30.0% and in France 28.0% of the total land area (Fig. 4.2).

There are differences regarding the geographical location of the forests:

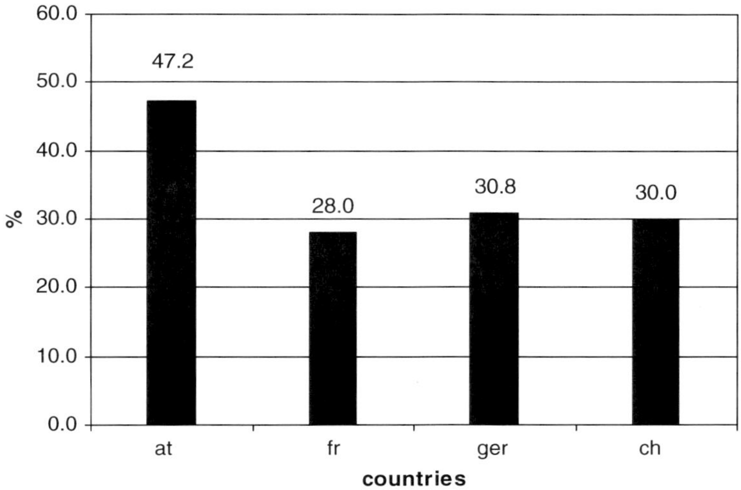

Fig. 4.2 Forest cover (%) in the Central countries

- In the Central countries most forests are found on mountainous areas where the soil is not productive enough for agricultural use.
- In Switzerland forest mostly dominates the mountain ranges of the Alps, yet 72% of the population can reach a forest within 20 min.
- In Austria, forest dominates in the alpine area and in the lower mountain ranges. There is less forest in the east of Austria because of agricultural use and in the high alpine areas because of the natural timberline and the alpine pastures. (BMLFUW 2005).
- In Germany, there is less forest cover in the northern part of the country. In this region there are lowlands that have always been used for intensive agriculture (Schleswig-Holstein, Niedersachsen, Mecklenburg-Vorpommern). Most forest cover can be found in areas where mountain ranges dominate (e.g. Spessart, Schwarzwald, Bayrischer Wald, Pfälzer Wald and the Alps). Generally, however, forests can be reached easily from nearly all parts of the country.
- In France there is also less forest cover in the northern part of the country. This region is characterized by lowlands and by the Atlantic climate. France forests are not spread evenly across the country, but two thirds of the population can reach forests easily for recreational use.

4.2.2 Development

In all Central countries there is a general trend towards reforestation by natural successional processes in areas where agriculture is no longer competitive. Particularly in the alpine area, mountain pastures are becoming abandoned and these areas become overgrown. This process also takes place in lower mountain ranges and other less productive areas.

4.2.3 Ownership

Ownership patterns differ widely in the central countries (Fig. 4.3). In Austria private forests clearly dominate: 82% of the forests are privately owned; 3% of the forests are in public and 15% in state ownership.

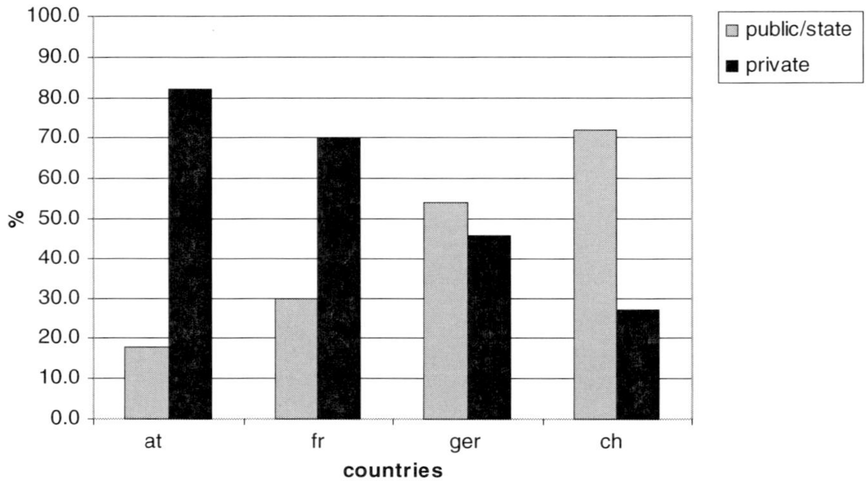

Fig. 4.3 Forest ownership in the Central countries (%)

In France, privately owned forests clearly dominate (70% of the total forest area). 18% of the forests are community owned and 12% are state owned forests. In Germany, slightly less than half of the forests are privately owned (46%). Twenty percent of the forests are community owned and 34% of the forests are state owned. In Switzerland only 27% of the forests are privately owned. Community owned forests clearly dominate (65% of the total forest area). The share of state owned forests is 7%.

In the Central countries there are clear differences with regards to ownership in many regions. In Germany private forests dominate in Bavaria and in the northern parts of the country (Nordrhein-Westfalen, Niedersachsen and Schleswig-Holstein) a lower proportion of private forests can be found in Rheinland-Pfalz, Saarland and in the federal states of the former DDR. There are historical reasons for this distribution of ownership patterns (Bundesministerium für Verbraucherschutz, Ernährung und Landwirtschaft 2005). In Switzerland, public forests prevail in the whole country but to a lesser extent in the midlands (Mittelland) and the foothills of the Alps (Voralpen). In France, public forests dominate (in ha) in the north-east regions (Lorraine, Alsace and Franche Comté) but are present all over the country. Private forests dominate in the western part of France (more than 90%).

In all Central countries the size of private holdings varies widely across the country. In Germany the private forest holdings are predominately small structured and split. Twelve percent of the holdings are larger than 1,000 ha, whereas 57% of the privately owned forests are held by small forest holders holding less than 20 ha. (Bundesministerium für Verbraucherschutz, Ernährung und Landwirtschaft 2005). In Bavaria, however, nearly 60% of the private forest area is owned by a few owners who hold more than 1,000 ha. On the other end of the scale a large amount of small forest holders (owning less than 50 ha) hold together only about 20% of the private forests (Bayrische Landesanstalt für Land- und Forstwirtschaft 2005). In Switzerland, the average size of the 3,300 state forest holdings is 270 ha. There are more than 240,000 private forest owners and the average size of private holdings is 1.3 ha (Buwal 2004). In France, the median size of the 1,500 state owned forests is approximately 925 ha and in the 11,000 community owned forests about 145 ha. The median size of the privately owned forests is 2.9 ha. One million private owners possess more than 1 ha. Fewer than 200,000 private forest owners (total number: 3.5 million) possess more than 10 ha. (Memento Afocel 2004; Les chiffres cles de la foret privee 2005). In Austria private forest holdings are generally rather small; half of the private forest is owned by nearly 170,000 owners with holdings smaller than 200 ha. One third of the Austrian forest is owned by large forest enterprises (BMLFUW 2005).

4.3 Legal Background, Public Access and Forest Functions

4.3.1 Legal Background

The legal background has a strong influence on recreational activities. In all Central countries there are laws affecting the recreational use of forests. In France and Switzerland there are general directions regarding recreation in forests. In Austria and Germany there are laws pertaining to recreational uses.

Legislation covers the aspects of public access and the collection of mushrooms, berries and minerals in all Central countries. In Austria, France and Germany sporting activities are also covered by legislation. There are also regional variations within all Central countries. In Germany, laws in the federal states are often different. In Switzerland there are differences in the federal laws at the level of the canton, e.g. concerning the collection of mushrooms. In France there are regional variations in community forests but not in state forests. In Austria there are regional variations regarding collecting activities, protected areas and tourism concerns.

4.3.2 Public Access

In all Central countries legislation enables forest owners to influence public access for recreational uses (e.g. recreational activities that do not require any special equipment such as hiking, collecting mushrooms and berries). Restrictions may be implemented in state and public forests in all Central countries; if nature conservation issues make it necessary, mainly minor access restrictions may be implemented.

- In France restrictions may also be implemented in private forests. Legislation allows owners to prevent public access.
- In Germany the aspect of public access is treated differently in the northern and southern part. For historical reasons there is public access to forest trails and roads in the Northern part; in the Southern part there are no limitations to public access at all.
- In Austria there is public access to forest roads and trails for recreational use without equipment. There are restrictions for recreational use requiring equipment like cycling, mountain biking, riding or camping. This is also a question of liability of the forest owner. The forest owner is liable for a favourable state of the forest roads and for accidents according to this favourable state of the roads. It is also forbidden to enter forest stands with trees less than 3 m in height.
- In Switzerland there is free access to the forests and there is no requirement to stay on forest roads or trails. Restrictions on equipment are possible but most activities are tolerated. Collection of berries, mushrooms, etc. is allowed, but can be restricted by local authorities.

The following table gives an overview of the legislation influencing public access and other recreational uses.

Table 4.1 Existence of legislation to influence public access and other recreational uses

	Austria	France	Germany	Switzerland
Existence of national legislation to influence public access (yes/no)	Yes	Yes	Yes	Yes
Restrictions for public access can be applied in:				
• Private forests	No	Yes	No	No
• Public forests	Yes	Yes	Yes	Yes
• State forests	Yes	Yes	Yes	Yes
Existence of laws affecting recreational uses:				
• Sport activities	Yes	Yes	Yes	No
• Collection of NTFP's	Yes	Yes	Yes	Yes
• Other uses	No	Yes	No	No

4.3.3 *Forest Functions*

Basically, forests have four main functions: recreation, timber production, protection of soil, water and avalanches and biodiversity protection. The recreational function is regionally important in the Central countries: near big cities and in tourist regions.

In Germany all functions are equal and have to be balanced. According to the legislation all functions have to be considered. Recreation is locally the primary function, especially in urban areas and in tourist regions (mostly mountainous areas like the Alps and the Black Forest). In France recreation is significant mainly in public forests, especially in suburban and urban areas and in tourist regions such as mountain and coastal areas. Recreation plays a less important role in private forests, as legislation allows owners to prevent public access. In Switzerland recreation is also important near urban areas and in tourist areas, but recreation has no priority at all at the federal administration level. Also in Austria, all forest functions have to be considered in the forest planning. In Austria timber production and tourism make the largest contribution to the GDP. The recreational function of the forest is guaranteed by the Austrian Forest Law. Recreation is locally more important in tourism and urban areas.

Compared to other European countries, Central Europe ranks at a middle position regarding the importance of the recreational function of forests.

4.4 Forest Managers' Education and Tasks Regarding Recreation in Forests

Knowledge about planning and management of forest based recreation is considered as part of the basic knowledge and skills of foresters in the Central countries. Figure 4.4 shows the evaluation of forest managers' knowledge about planning and management of forest based recreation.

In Germany and Austria this basic knowledge is acquired mostly during education and completed through working experience. In Germany both education and working experience are considered as good. In Austria education in recreation is considered to be average and working experience as good. In Switzerland the level of both education and the value of working experience regarding recreation are considered as rather unsatisfactory. In France, the education is considered as rather unsatisfactory, but the working experience is considered as good (here short post-educational training courses may play an important role and may be considered as working experience). Over the last twenty years, there has been a training system called FOGEFOR (formation à la gestion forestière) which is open to all forests owners. They learn how to manage their forest from this system. Some of training session are specialised on topics, of which recreation could be one.

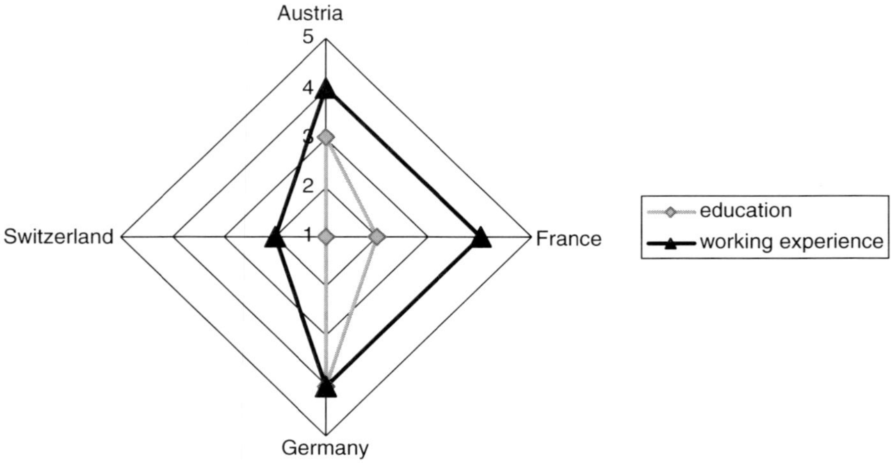

Fig. 4.4 Evaluation of knowledge about planning and management of forest based recreation obtained by education and working experience (1=unsatisfactory, 5=excellent)

The reactions from the Central countries to the question about tasks regarding recreation in forests clearly show the differences in their organisational systems, even within the same country.

Concerning the tasks of the forestry administration, all countries except France mainly consider planning as their main task. In France (and this seems because of the still centralised organisation) the main task of the forestry administration concerns policy, while planning is the responsibility of forestry management at the regional level.

In all the countries planning is also one of the main tasks of forestry management. In Switzerland the main task is information collection followed by planning, supervising and establishing infrastructure and environmental education. In Austria, France and Germany planning, information and environmental education, management actions for visitor guidance supervising and establishing infrastructure monitoring as main tasks of the forestry management. In France an other task is mentioned that concerns structured public consultation in the way that public opinion is more and more taken into account in actions concerning planning and management especially in public forests and not only for recreation matters.

4.5 Conflicts

4.5.1 Types of Conflicts

Conflicts affecting recreation in forests can be subdivided into different types. There may be conflicts between different recreational groups (e.g. horse-riders – cyclists, cyclists – hikers) or conflicts due to overcrowding. Conflicts may arise between

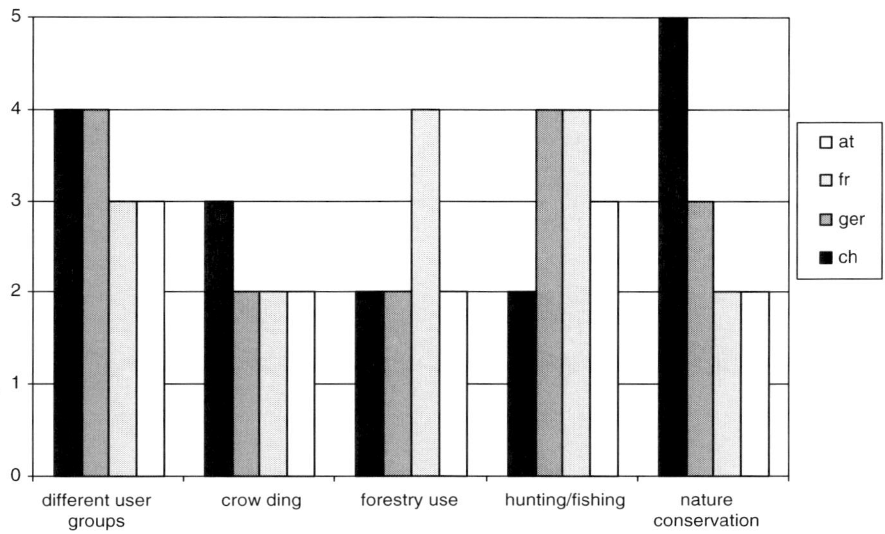

Fig. 4.5 Conflicts levels in the Central countries (1=not important at all, 5=very important)

different land uses, e.g. timber production and recreational use (e.g. due to safety issues or aesthetic impacts) or conflicts between recreation and other uses in the forest (e.g. hunting or fishing). There may also be conflicts between recreation and nature conservation issues.

The main conflicts in the Central countries arise between different recreational groups, out of collecting activities, from traffic noise and road construction (Fig. 4.5).

In Germany and Austria the most important set of conflicts arise between different recreational groups; another serious problem in Germany is hunting; in Austria, due to hunting locally important conflict arise. In France the most important conflict is between recreation and forestry use, while in Switzerland the most important conflicts are between recreation and nature conservation and between different recreational groups.

In general, conflicts between different recreational groups are of the greater importance in the central region, especially in Germany and Switzerland, though less so in France and Austria. Typical examples are conflicts between hikers and mountain bikers and between dog owners and joggers.

Conflicts between recreation and nature conservation are likewise important in most parts of the central region. In Germany, conflicts exist mainly in areas with greater restrictions for nature conservation (e.g. in national parks and nature parks) and they predominantly involve outdoor sports. In Switzerland, conflicts with nature conservation concern urban as well as rural and Alpine recreational forests. This type of conflict seems to be of somewhat lower significance in France and Austria.

Conflicts between recreation and hunting/fishing, however, are quite important in France. The other countries report a medium level of conflict between recreation and hunting. Locally this conflict is very important in Austria. Recreational activities often affect hunters.

Crowding effects are generally only of medium importance in the central region and occur locally in recreation hot spots (in Germany) and mainly in urban forests (Switzerland and Austria).

Conflicts between recreation and forestry use seem to be of high importance solely in France and only of medium or low importance in Switzerland, Austria and Germany. However, current results from a survey among Swiss foresters indicate that the problems between recreation and forestry use are underestimated. Foresters often report severe interruptions to their work, because forest visitors ignore road closures and recreational activities can incur additional costs for measures necessary to guarantee the security of forest visitors. Moreover, foresters report a lack of understanding about forestry use. Forest visitors are unwilling to accept constraints due to timber production. In the perception of Swiss foresters, this often leads to conflicts between forestry use and recreational activities.

4.5.2 Important Activities and Their Conflicts with Nature Interests

In the Central region recreational activities that lead to the most frequent conflicts with nature interests in forests are mountain biking, collecting activities, picnicking, skiing and ski touring. The following chart shows the frequency of conflicts for different recreational activities (Fig. 4.6).

Conflicts concerning winter sports such as alpine skiing, ski touring, snowshoeing and cross-country skiing are very frequent in all Alpine and sub-Alpine areas of Switzerland and Austria – partly also in Germany (some Alpine areas and the Harz Mountains) and to some extent in France (Alps and Pyrenees). Snowmobiling as a recreational activity does not exist in the central region.

Severe conflicts concerning the natural environment often emerge because of these activities, mainly in Switzerland and Austria (Alpine skiing, snowshoeing, ski touring) as well as in Germany (particularly ski touring). Alpine skiing outside marked slopes often affects wildlife (disturbance) and protection forests (damages to trees) negatively. Now, to secure the snow during the winter there are big artificial lakes to provide water for artificial snow making leading to hydrological disturbance in the watershed.

By contrast with ski touring, snowshoeing is very easy to learn and does not require any special skills. The frequency of this activity has increased rapidly in Switzerland during the last five years and a future rise in the number of snowshoe walkers is likely. Therefore, it is anticipated that ecological conflicts will increase and that effective guidance tools are necessary. Snowshoeing is often carried out in remote areas, where rare species live (such as capercaillie). Because of this, the severity of conflict is rated as high in Switzerland. In Austria the severity is rated

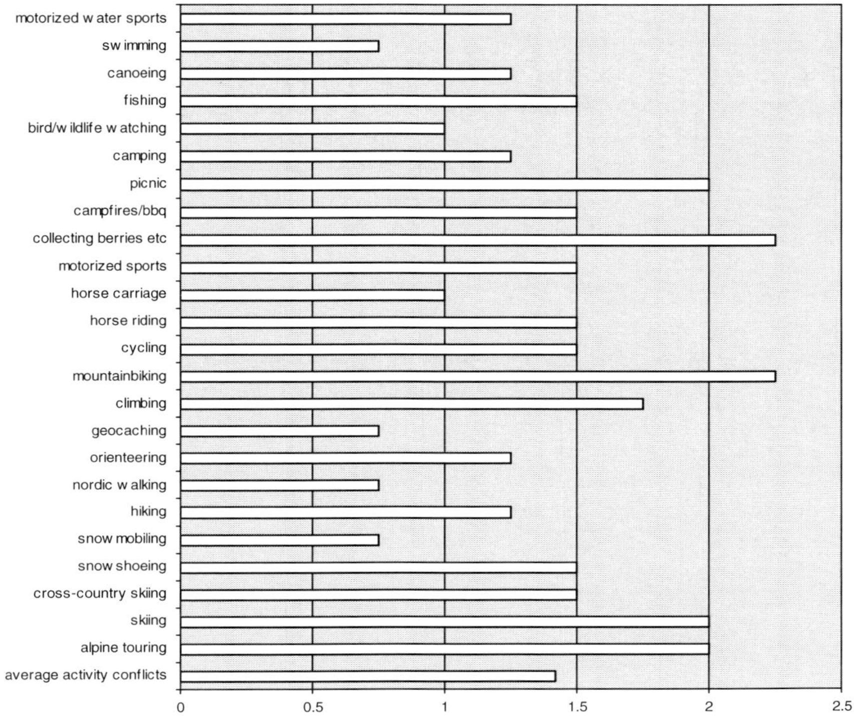

Fig. 4.6 Frequency of conflicts with recreational activities (0=non-existent, 1=rarely, 2=medium, 3=frequently)

medium. In the other parts of the central region, conflicts regarding snowshoeing are not mentioned though in Germany an increasing use has been reported from mountainous and alpine areas.

Hiking and nordic walking are popular activities in Austria, Switzerland and Germany, but less common in France. Both activities rarely lead to conflicts, as hikers and walkers generally keep to footpaths and hiking trails. Orienteering and geo-caching is done in the whole central region, but not very frequently. The potential for conflicts is higher than that of hiking or nordic walking, because leaving footpaths and going straight through the forest is a necessary part of both activities. This could cause trampling in areas otherwise undisturbed by the public and may frighten wildlife in its refuge. As orienteering is mainly practiced by clubs or other associations, there have been several contacts between some of these associations and the forestry service in Switzerland and as a result, associations have started to mark wildlife refuges on their maps and ask members not to enter these areas.

Climbing is a very important activity in all four countries, and in Germany and Switzerland conflicts with the natural environment are frequent. Climbing on cliffs below the timber line damages the specialized vegetation, favouring the survival of plants more resistant to trampling, but not necessarily specialized to cliffs. In the Jura Mountains, the problem is accentuated by the numerous endangered species found on these cliffs - mainly glacial relic species using these cliffs as a last refuge. Climbing also disturbs birds nesting on the cliffs. Models for contracts and voluntary agreements exist in several climbing sites in Germany, some of them with satisfying results (BfN 2004).

The most relevant recreational activity in forests of the central region involving a means of transport is clearly mountain biking. Ecological conflicts are quite frequent and their severity is rated as high in Germany and Switzerland and as medium in Austria and France. Mountain biking is very popular in the whole region and is carried out in any kind of forest (ranging from urban forests to Alpine forests). Mountain bikers often choose and prefer small paths and single trails that have originally been used only by hikers. Furthermore, some of them even create new paths by criss-crossing through the forest (downhill biking) or building obstacles such as jumps or drops (freestyle biking). All these activities may lead to disturbances of fauna and flora in the forest. However, the type of damage is not clearly specific to any country. To solve problems or to reduce impacts special slopes for downhill biking have been established. Here often cooperation between tourism organisations and forestry was needed. In Austria mountain biking is restricted to approved trails for liability reasons. The severity of ecological conflicts caused by cycling is rated as medium, as is horse riding; horse carriage driving is even less relevant.

Motorized sport is generally forbidden in forests of the central region. In spite of this common situation, results vary between the three countries: France does not report any frequency or severity of conflicts, Switzerland rates both dimensions as low, Austria medium and Germany indicates severity and frequency of ecological conflicts caused by motorized sport as high.

Collecting berries, mushrooms, minerals, etc. is an important activity in all three countries and conflicts regarding the natural environment and nature conservation are reported to be frequent. Conflicts can arise from gatherers leaving footpaths to find berries, etc. as well as by over-exploitation of resources. Therefore, these activities are partly restricted and controlled in Germany and Austria, while mushroom picking is limited to a certain amount per person and is only allowed on certain days (e.g. during one day a week) or during a specified time period in Switzerland. In the south of France, there are conflicts between forest owners and especially professional mushroom pickers. There are different forms of authorisation; new solutions like a "mushrooms path" have also been developed. The principle is to pay for the right to pick mushrooms during the season.

Picnicking and enjoying campfires is popular in the central region. This activity leads to several conflicts. Areas surrounding the campfires can be badly affected by trampling, live branches are cut to make skewers and the collection of fuel wood depletes the woody debris in the forest. The problem is accentuated in Switzerland because of the lack of laws regarding campfires, whereas in forests in Austria, Germany and France campfires are only allowed in official sites.

Picnicking plays a particular role in Germany and Austria, but not in France. Camping is not a very important activity in the forests of the central countries. In

Germany and Austria camping outside official sites is only possible with the agreement of the land owner. Conflicts due to picnicking are frequent in the whole central region, including France, while conflicts due to camping are not a big problem. Wildlife and bird watching are of moderate importance in the central region and so are the conflicts accompanying these activities.

All activities in connection with open water such as fishing, swimming and canoeing are of moderate or even minor importance in the recreational forests of the central region. The most problematic activity is enjoying motorized water sports. The use of motor boats for boating, wakeboarding, water skiing, etc. too close to lakeshores can scare away waterfowl and disturb birds breeding among the reeds. This is however not a problem specific to forests, but concerns lakes in general. Small forest lakes are usually relatively natural thanks to their often remote location. Motorized sports are rarely possible here, but such lakes are very attractive for swimming and fishing. As a habitat for all species dependent on a forest-water ecotone, for example amphibians and in some cases vegetation typical to fens, these lakes have an important ecological function and sometimes are part of a nature reserve. Therefore, people swimming can cause some disturbance to these otherwise undisturbed lakes and previously untouched areas along the shore containing rare moor vegetation can get trampled. In Switzerland fishing may be allowed in forest lakes, but as a rule there are regulations concerning the fishing license, number and size of fish and equipment allowed.

Several countries mention another recreational activity, which is – despite, or maybe because of its obvious relevance – often neglected: dog walking. The frequency of this conflict is rated as medium and the severity as high to medium. Current survey data show that Swiss foresters rate dog walkers as one of the main ecological conflicts in Swiss forests. Forest visitors often fail to put their dog on the leash, which can lead to dogs chasing and killing game. Furthermore, owners frequently leave the dogs' faeces in the forest, which annoys other visitors, especially in Switzerland, where special bins and small plastic bags are provided at forest entrances for owners to pick up the mess. Dog owners are also often involved in social conflicts because other visitors are scared of dogs running loose or because they are disturbed by dogs barking. In Austria and Germany these conflicts are also relevant especially in urban forests.

4.5.3 Impacts and Disturbance of Other Societal Activities on Recreation

All four countries of the central region report conflicting impacts from other uses with forest based recreation and nature tourism. Figure 4.7 shows the frequency of impacts on forest recreation.

Traffic noise is clearly a problem in the whole central region. Frequency and severity are always rated as medium or high. Aircraft noise also causes some impact, but only in specific areas (such as around the airport). Military noise is

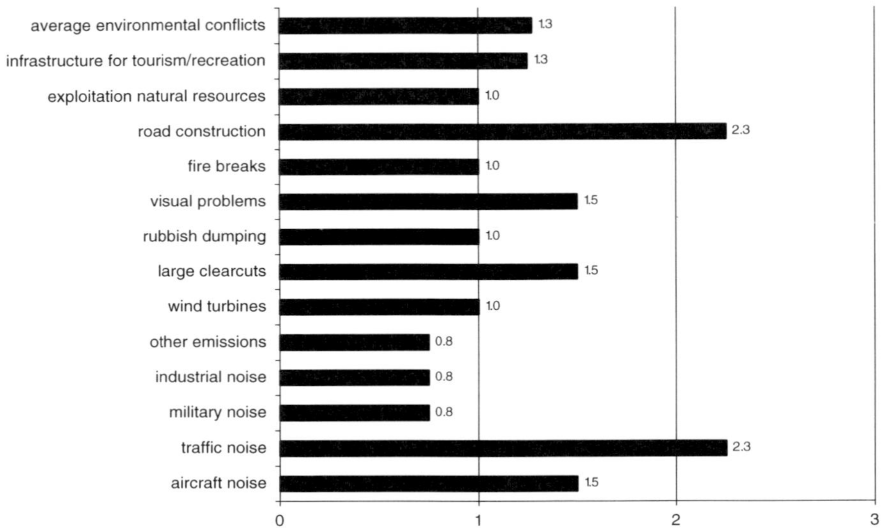

Fig. 4.7 Frequency of impacts in the Central countries (0=non-existent, 1=rarely, 2=medium, 3=frequently)

rare, but the subsequent conflicts are rated severe in all central countries except Austria. Industrial noise is rare and does not cause any severe conflicts. This may be explained by the fact that there are few industrial zones in the recreational forests of the central region.

Conflicts in connection with wind turbines are mentioned in Germany. In France there are also more and more wind turbines in forest areas; people are rather divided concerning their acceptance but there are few studies yet on this topic. In Switzerland wind turbines are infrequent. In Austria they pose hardly any conflicts in forests.

Negative impacts caused by large clearcuts are rare in Switzerland in Germany, but rated as rather severe. However, the relative extent of clearcuts in Switzerland is small: compared to countries like Scandinavia or Scotland there are no true clearcuts. In Austria the impact of large clearcuts is rated as not very severe but rather frequent. France indicates both dimensions – frequency and severity – on the medium level and states that the degree of conflict depends on the type of the forest area.

The frequency of conflicts caused by rubbish dumping is low in the central region, but increasing in Germany. In Germany there are large differences between rural and urban areas. In urban areas the severity is high whereas in rural areas littering is not such a problem. The severity of this impact is high in France.

There are some conflicts caused by visual problems of damaged forests in Switzerland and in some regions in France. Visual problems in Switzerland are due to a storm in 1999 that caused damage in several areas and affected the appearance of some forests. Some forest visitors also tend to criticise forestry use of dead wood as "forest damage" and impairment of the beauty of a forest. Damage caused by storms also occurred in several parts of Germany (in a large scale for example in Black Forest). While this is accepted as a natural event there were intensive discussions concerning damage by beetles and other insects in protected areas. There was also a discussion on the "need for chemical help" for the forest not only by land owners but also by the general public. In most cases only information was provided by the forest administration.

Conflicts caused by road construction are frequent and severe in all countries of the central region. Infrastructure for tourism and, in France, electricity transmission lines, often leads to severe conflicts. For example, infrastructure for winter sports, such as ski-lifts, which are essential for winter tourism, often badly impair the landscape of Swiss and Austrian tourist centres. Damage becomes obvious during the summer season and clashes with the aesthetic expectations of tourists. Of course there are also ecological problems linked to the extensive landscape impact of infrastructure – especially if this is situated in remote and undisturbed areas.

4.6 Planning, Management and Monitoring

4.6.1 Planning for Recreation and Nature Tourism

In all countries of the Central region there is independent planning for recreation and nature tourism in forests. This planning does not take place at regular intervals but is only done in special cases (Table 4.2).

Table 4.2 Availability of independent planning (legally required or optional) in the Central countries

	Austria	France	Germany	Switzerland
Independent planning for recreation and nature tourism in forests	Yes, in special cases	Yes, in special cases	Yes, in special cases	Yes, in special cases
Legal or optional requirement for forest owners	Optional	Legally required (state/public) Optional (private)	Optional	Optional

In Germany and Switzerland the planning is part of the general land use or landscape planning. The planning is integrated in forest inventory or management planning in France and in Germany, partly in Austria in large and state owned companies. In Austria there is recreation-related planning on special topics like mountain biking and interpretive trails. In France the planning is legally required only in state and public forests; in private forests it is optional. In Austria, Switzerland and Germany this planning is optional.

In Austria, Germany and France there are handbooks, guidelines ore other instructions or frameworks to direct the planning and the management of recreation and nature based tourism. In France there is a national guide for forest management as well as guidelines for recreation, landscape and cultural heritage management; in Germany there are guidelines concerning mapping of forest functions (Waldfunktionsplanung), forest inventory (every 10 years) and landscape planning. In Austria there are handbooks for pilot projects concerning forest and tourism. In Switzerland there are no instructions or frameworks to direct the planning or management of recreation and nature tourism in forests.

4.6.2 Planning Steps

The planning steps that all Central countries undertake when doing recreation related planning includes definition of vision and goals and public participation at that stage (Table 4.3).

Table 4.3 Steps that are undertaken for planning forest recreation at regular intervals in the Central countries

Steps in planning forest recreation	Countries
Inventory of ecological conditions	FR, GER
Inventory of social conditions	FR, GER
Analysis of conflicts	AT, GER
Definition of vision and goals	AT, FR, GER, CH
Participation in the definition of goals	FR, GER, CH
Definition of standards	AT, GER
Deduction of measures	AT, FR, GER
Participation in the deduction of measures	FR
Monitoring	FR, GER

In France and Germany an inventory of the social and ecological conditions is carried out in planning for recreation and nature tourism in forests as well as definitions of prescriptions and monitoring. Planning steps that are carried out only in Germany are analysis of conflicts and definition of standards. In France there is also participation in the definition of prescriptions and financial partnerships.

In all Central countries land owners and community representatives regularly take part in public participation activities and events. In Germany, Austria and Switzerland land users, sport associations and hunting and fishing representatives are included. In Austria and Switzerland local residents are also included. In France local societies take part in participation.

In Germany and Austria monitoring is done less than once in 5 years and in France at irregular intervals or whenever necessary.

In the Central countries the main sources of data available for recreational planning are regional surveys. Data collected by the forest administration are available in Austria, Germany and Switzerland. Data collected by other institutions for the planning process are partly available in Austria, Germany and France but only for intensively used areas or areas where there are conflicts (e.g. nature conservation). Data of mostly national surveys are available in Switzerland.

4.6.3 Overall Evaluation of Recreation and Tourism in Planning and Monitoring

The Central country experts evaluated the overall state of development of forest-based recreation and nature tourism in forests, considering ownership structure, the legal situation and the education of management personnel on a five-point scale. France and Switzerland deem the consideration of forest based recreation as "below average", Austria as "average" and Germany as "good".

Looking at regional differences the results turn out better: Germany rates the way that forest recreation around big cities and in tourist regions is considered as "excellent". Austria and France rate regional development as "good". In Austria there are regional differences associated with urban or tourism areas and rural areas. In France the consideration is locally better due to pressure by politics and public and the availability of public forests (access). Switzerland rates the way that forest based recreation is considered regionally as "average" due to the interest and motivation of foresters.

4.7 Recreational Infrastructure

Generally, as the basic infrastructure for recreational use of forests mainly consists of forest roads and trails, all Central countries consider that the supply is "good" to "excellent" (see Fig. 4.8). This opinion applies not only to urban areas but also rural and remote areas. In Austria there are standards for the provision of access: in urban areas 70 running metres/ha, in rural areas 50 running metres/ha, in urban areas 30 running metres/ha. There are differences in different federal states, in Salzburg the

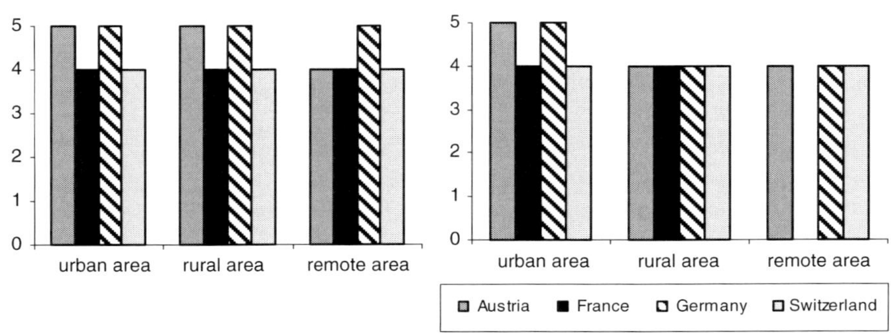

Fig. 4.8 Supply of basic infrastructure (*left figure*, **a**) and additional infrastructure (*right figure*, **b**) (no bar = non-existent/no data, 1=unsatisfactory, 5=very good)

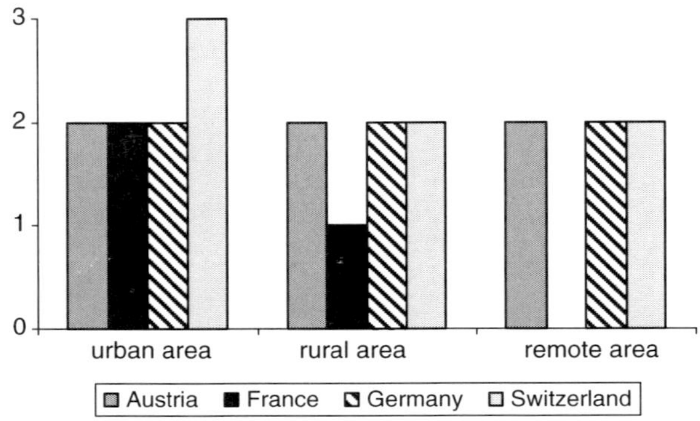

Fig. 4.9 Quality of infrastructure for recreation and nature-based tourism (no bar = non-existent, 1=unsatisfactory, 3=excellent)

supply of basic infrastructure is slightly higher: in urban areas 80 running metres/ha, in rural areas 60 running metres/ha and in remote areas 40 running metres/ha.

The amount of additional infrastructure seems to be higher in Germany and in Austria than in Switzerland and in France, both in urban and rural areas. In France this infrastructure is quite absent in remote areas, while it is significantly present in the other Central countries. Figure 4.8 shows the supply of basic and additional infrastructure in the Central countries.

Concerning the quality of the infrastructure, all Central countries consider it "excellent" to "good" in all areas, except some rural areas in France where the quality is considered as "unsatisfactory" (see Fig. 4.9). This opinion refers on one hand to the incorrect choice of materials and on the other hand to a lack of maintenance of several types of infrastructure. Other types of infrastructure picnic tables, toilets and also "health circuits" are often constructed of wood. To develop some activities of climbing, there is the concept of "Go Ape" or the organisation of a climbing circuit in the trees. At present in France, there are more than 200 centres, most in the alpine region, to complete with other alpine activities.

Additional recreational infrastructure for forest recreation and nature based tourism is more frequent in more central areas. The most common recreation infrastructure includes benches, playgrounds, picnic areas, rubbish bins, look outs, nature trails and infrastructure for environmental education. Figure 4.10 shows the occurrence of additional infrastructure in urban, rural and remote areas in the central countries.

In Switzerland forests are not rated as over-equipped for experiencing nature. As well in urban, rural and in remote areas they are considered to be somewhat under-equipped. In Germany and Austria there are different opinions on the topic: however it seems that, especially in urban areas, forests are rated as being over-equipped. In France, urban and rural areas are rated as being over-equipped whereas remote areas are considered to be under-equipped.

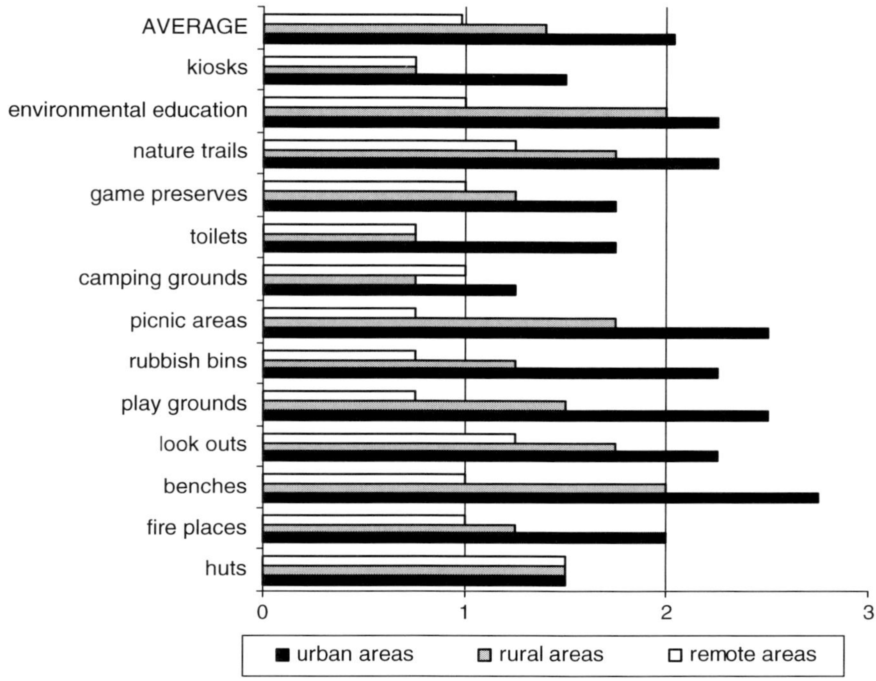

Fig 4.10 Occurrence of additional infrastructure in urban, rural and remote areas in the Central countries (0=non-existent, 1= rarely, 2=reasonable, 3=frequently)

4.8 Approaches to Visitor Management

The most common actions for visitor management can be divided into three types: restrictive management actions, soft management actions, and facilitating management actions. Although it is difficult to compare management actions from one country to another, it is interesting to note differences concerning both the actions and their acceptance. These observations can give birth to renewed approaches. The following figure and table shows the frequency of use of management actions in the Central region in urban, rural and remote areas and the level of acceptance of these actions.

4.8.1 Restrictive Management Actions

In the Central region, visitor management is mainly implemented by restrictive actions such as applying rules, regulations and prohibitions as well as marking protected areas and taboo zones, especially in urban areas (see Fig. 4.11). Marking protected areas is the most common tool for visitor management in Germany, while

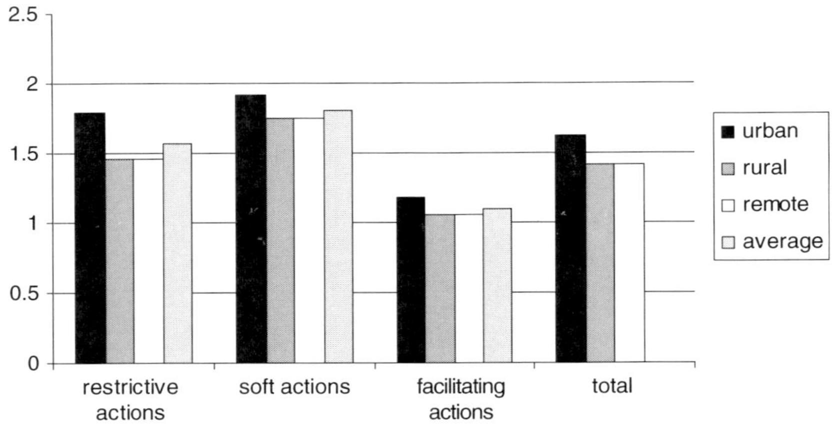

Fig. 4.11 Occurrence of management actions in the Central region, in general and in urban, rural and remote areas (0=non-existent, 1= rarely, 2=regularly, 3=frequently)

this tool is used less frequently in Switzerland. Entrance fees aimed at reducing the number of visitors and at covering the costs of recreational activities are rarely charged in the Central countries. The acceptance of entrance fees is rather low, which may be a reason why there are used rather rarely in the Central countries. Of all the direct management actions, marking of protected areas and taboo zones is accepted best by the public, followed by rules and regulations and prohibitions. In other words, the most frequently implemented management actions are also the ones accepted most readily.

Among the most unpopular actions are the charging of entrance fees, so this action is rarely or never used (yet) directly (see Table 4.4). In Austria it is possible to charge entrance fees for commercial events like running events or mountain bike

Table 4.4 The level of acceptance in the Central region for the various management actions (high >= 2.5, medium > 1.5 and < 2.5, low =<1.5)

Acceptance by the public	Restrictive management actions	Soft management actions	Facilitating management actions
High	Marking protected areas/taboo-zones	Attractive infrastructure Educational infrastructure	Guiding thematic tours Guiding ranger systems
Medium	Rules and regulations Prohibitions	Signposting responsible visitor behaviour	Training special user groups Training commercial providers
Low	Entrance fee Barriers and fences Limiting visitors' entrance		

races. Local, regional or national taxes are a more indirect way used in some regions or countries. Public opinion could however be prepared to the principle of entrance fees if incentives and facilitating actions can reasonably justify it. For example, the climbing circuit in the trees is accepted in France as being similar to any other a leisure activity, and more than 1.6 million people per years pay an entrance fee around 15 Euros each. The entrance fees were accepted as soon as the service was provided.

The acceptance of restrictive actions could possibly be higher, if correctly presented, designed and integrated in the landscape: most cases of low acceptance of management actions are often due to a lack of explanation or lack of social demand.

4.8.2 Soft Management Actions

In terms of incentive management actions, improving areas with new attractive infrastructure is the most common action. It is used in rural areas slightly more frequently than in urban and remote areas. Especially in urban areas, new infrastructure can also be installed for the purpose of ecological education. Sometimes signposts are put up, asking visitors to act in a responsible way. The acceptance of new infrastructure is high, probably because it benefits the users directly, and signposts are also moderately well accepted. Visitors are rarely guided by rangers, possibly because this is a rather expensive measure. However, the use of a ranger system is very well accepted by the public, so there seems to be a potential for this system to be used more frequently. Limiting visitors' entrance is used only rarely in all countries. In Germany and Switzerland the acceptance of this management action is low and in France and Austria medium.

However, examples from North America indicate that limiting numbers of visitors entering a place might be necessary in certain cases for long-term nature protection. With a clear explanation, the degree of acceptance of this can be improved. To make this management action more acceptable, other measures and infrastructures can be provided simultaneously to counterbalance the negative effects or impacts. As there are lots of types of public, it occurs difficult to satisfy every one at the same time at the same place by incentive actions.

4.8.3 Facilitating Management Actions

Facilitating management actions are of low importance in the central region. On some occasions, mainly in urban areas and mainly in Switzerland, certain user groups have received special training, but the training of commercial providers has been neglected up to now. Thematic tours for visitors are rarely or occasionally offered in most central countries but the acceptance of such tours is high. Like the ranger system, there is a potential to do more in this field and have a direct interaction between recreation seekers and rangers or people responsible for the recreation areas.

In France especially, landscape management actions have to be taken into account. This is action that concerns design as well as silviculture, aiming to provide a diversified close-to-nature environment in which a large range of the public can find a good reason not only to visit a place, but also to create a sense of loyalty to it. Generally, visitor management is more important in urban than in rural or remote areas with more tools being made use of more frequently. This is most likely due to the high number of visitors in urban forests, making visitor management necessary in the first place. The fact that some management actions remain very controversial among the public in different Central countries, makes it clear that these actions have to be seen in their social and environmental context.

4.9 Future Trends and Challenges

Considering future trends and challenges, three aspects should be discussed:

- Trends and challenges concerning forest and landscape development;
- Trends and challenges caused by new recreational activities and changes in the behaviour of visitors;
- Challenges for science and research.

Overall it has to be recognised that for most people in the Central countries the recreational functions of forests is of increasing importance. The production function is by comparison of lower interest and has lower awareness. With the trend to a more flexible working pattern, more short-time holidays and the enlargement of settlements in agglomerations the challenges for forestry in densely populated areas will increase. It should be debated whether in those areas a special concept for recreational purposes should be the standard (as opposed to being only occasionally done). Here the Central countries could profit from the experience from several Nordic countries especially Finland. In all Central countries and in particular in the mountainous areas a retreat from traditional land use due to soil conditions and an increasing amount of forest cover, mainly by natural succession, can be seen. This has effects on the character of the landscape and its suitability for tourism and recreation. Challenges for the future are here to find the balance and accepted trade-off between the maintenance of traditional land use, biodiversity and nature conservation as well as aesthetic beauty adapted to the different types of landscape. Studies from Hunziker (2000) and Lindenau (2002) show that many people would accept a higher proportion of forested land. There is the need for more research in this field.

The Central countries are characterized by the idea of free access and a high importance of forests as an area for leisure and sport activities. In Germany this is included in legislation. The trends in sport and recreation show an increasing diversity of activity types which are often followed by new forms of impact and threat. Thus the trend to more individualistic forms of sport activities such as snowshoeing leads to more disturbance and deterioration of wild life habitats. Therefore the challenges for forestry and science are to follow those trends and to provide and to share to experience of good practice.

In the Central countries the forests do not only play an increasing role for recreation but also for nature conservation. In most of the Central countries the majority of the sites belonging to the European network Natura 2000 or other nature conservation areas are located in forests. Therefore the balance between forest use, nature conservation and recreation will remain a challenge for the future. Here support from research concerning habitat deterioration, management planning (considering both recreation and conservation) and monitoring are needed. In addition, there are further challenges for research into visitor management and behaviour.

In order to progress approaches for visitor management, it seems obvious that the sociological and even ethnological approaches will have to be complemented by a psychological and ethological approach in order to understand more about informal behaviour of people in a "natural environment". This concerns not only the restrictive actions, but even more the incentive and facilitating actions. In France and probably also in other European countries this approach has not really been applied to forest recreation. The ethological approach is in common use in the world of marketing and especially in publicity and advertising. In France this approach has been experienced particularly in urban forest areas. Some of the results allow forest managers to make an area more attractive for visitors by very simple actions (Breman 2004).

Chapter 5
Mediterranean Region

George Pattichis, Savvas Kazafaniotis, Maro Kazepi, Athanasios Hatzistathis, Jose Manuel Castro, Goran Rubin, Barbara Mariotti, and Ljiljana Comic

5.1 Introduction

The Mediterranean countries participating in the COST Action E33 are: Cyprus, Greece, Croatia, Portugal, Italy, Serbia and Bosnia-Herzegovina (Fig. 5.1). Geographically, these countries are distributed from the eastern Mediterranean area to the coasts of the Atlantic Ocean. Parts of France, which is treated as one of the central European countries has also parts with Mediterranean character and similar

Fig. 5.1 Countries within the Mediterranean region

G. Pattichis (✉)
Department of Forests, Ministry of Agriculture, Natural Resources and Environment, 1414 Nicosia, Cyprus
e-mail: management@fd.moa.gov.cy

U. Pröbstl et al. (eds.), *Management of Recreation and Nature Based Tourism in European Forests*, DOI 10.1007/978-3-642-03145-8_5,

features to the other countries discussed in this chapter. Spain was not part of the Cost Action, so that there is no data available.

A distinctive characteristic of the Mediterranean region is the long and dry summer with maximum temperatures reaching 40–44°C in some countries and the uneven distribution of rainfall during the year. Also, rainfall varies considerably within the same country. For example, in Cyprus the mean rainfall in the central plains is 350 mm, whereas in the high mountains it is 1,000 mm and in Italy rainfall can vary from 500 mm to 3,000 mm.

5.1.1 Vegetation – Forest Cover

Due to the great variations in temperature and rainfall the vegetation cover varies considerably within a single country as well as across the region. Forests comprise both natural woodlands and plantations, the majority being natural forests growing mainly on the hilly and mountainous areas of these countries. The forests located in the plains and flat coastal areas can be locally very important for economic, social and recreation purposes (for example the pinewoods along the Italian, Portuguese and Croatian coastal belt and islands, the cork-oak woods in Portugal or the evergreen oak stands in Italy).

A characteristic of the Mediterranean landscapes is the presence of pre/post-forest shrubby vegetation influenced by various factors (low rainfall pattern, recurrence of fires, land misuse, etc.), mainly dominated by scleroplhyllus and evergreen species forming the so-called maquis-like ecosystems. The relative importance of maquis-like vegetation on the total vegetation cover varies greatly between the countries.

Another characteristic is the richness in plant species. For example in Italy there are about 5,400 species, in Greece more than 5,000 species and in Cyprus, with an area of only 9,251 km^2, there are almost 2,000 plant species. Also endemism is very high due to the variation of altitude, rainfall, geology, (micro)climate, etc. In Italy, for example, there are 712 endemic species.

Generally, high forest is the main variety in all countries, but a considerable part also comprises low forest (maquis) and other wooded land.

5.1.2 Forests and Tourism

The economy of the countries Mediterranean countries depends greatly on tourism and forests play a very important role towards this direction. Picnicking in forests seems to be the most popular type of recreation in all countries, particularly during the summer time (June–September). Thus, all countries put a strong emphasis on the development of picnicking facilities. Hunting is also a very popular activity and forest areas play an important role on this type of recreation. Some countries have also developed infrastructure for special interest recreational or tourism activities in forest areas such as nature trails (the European long distance path E4 and other thematic nature trails), cycling routes, thematic routes (for example wine routes, or religious route such as Byzantine Cyprus).

5.1.3 Forest Threats

A common phenomenon for all Mediterranean countries is the frequency and recurrence of forest and scrub fires, which cause severe damage to the forest and threaten all kinds of recreation facilities. Therefore, the design of certain types of recreation facilities, such as picnic sites or camp sites, needs special skills and knowledge in order to be safe on the one hand and offer recreational pleasure on the other.

In the Mediterranean countries, due to the density of recreation types and numbers of people taking part, nature conservation is in many cases at risk. This is particularly the case in mountainous areas, where the majority of recreation facilities are located; ensuring the conservation of nature is one of the challenges that are faced and its solution is quite difficult and costly.

Furthermore, maintenance of recreation sites and facilities is frequently rather poor and there is criticism from the public. Finally, it can be said that due to the high demand for forest recreation and the available funds, more effort must be made by the various authorities involved in order to offer reasonably good quality recreation facilities.

5.2 Forest Cover, Development and Ownership

5.2.1 Forest Cover

The variation of forest cover among the Mediterranean countries is between 26.7 and 53.4% and the mean is 38.8%. Bosnia, with 53.4% has the highest cover and Serbia, with 26.7% has the lowest forest cover (Fig. 5.2).

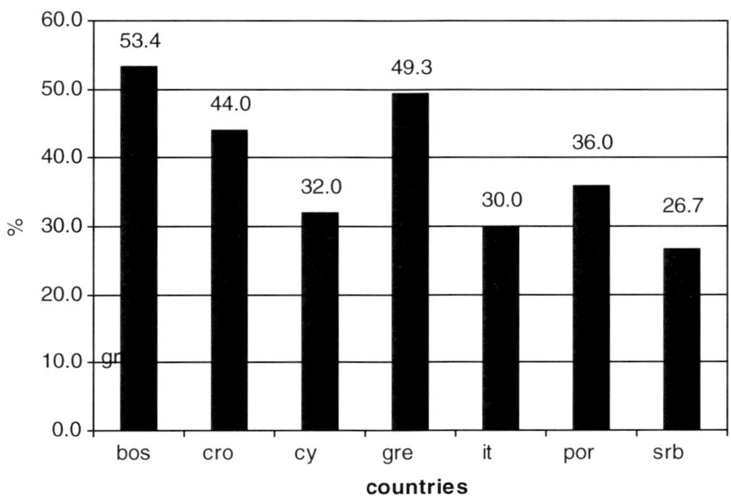

Fig. 5.2 Forest cover (%) in the Mediterranean countries

5.2.2 Distribution of Forests

In most Mediterranean countries, forests are not spread evenly but 2/3 of the population can usually easily reach forests for recreational use. Greece forms an exception: only 1/3 of the population can reach forests easily because the forests are mainly located in mountainous areas which are at a distance from the main population centres.

5.2.3 Forest Ownership

Forest ownership is one of the main factors influencing forest based recreation. As can be seen in Fig. 5.3 in the majority of the Mediterranean countries (Greece, Bosnia, Croatia, Cyprus, Serbia), the forests are generally publicly owned (state and communal) whereas in Portugal over 80% and in Italy 66% of forests are privately owned, although in Cyprus and Serbia there are also quite large proportions of private forest.

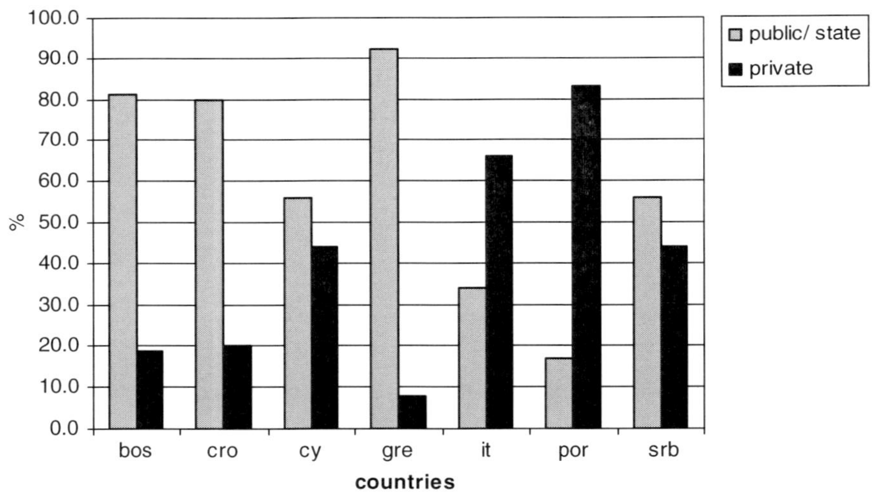

Fig. 5.3 Forest ownership in the Mediterranean countries (%)

5.2.4 Regional Differences

Portugal and Italy also show clear regional differences and, because of the high percentage of private forests, the size of holdings varies widely across the country. For Cyprus, Bosnia and Croatia there are some regional differences regarding the ownership and size of private forests but only in a few areas. For Greece and Serbia there are no differences.

5.3 Legal Background, Public Access and Forest Functions

5.3.1 Legal Background of Public Access

In all the Mediterranean countries legislation enables forest owners to influence public access for recreational uses. In Greece no legislation exists. Table 5.1 shows sectors where restrictions may be implemented in each country. Access is partly restricted in most Mediterranean countries through the use of barriers, limiting visitors' entrance, prevention of access etc. In Greece, Italy and Cyprus, for example, access is restricted under particular circumstances (such as National Parks, Aesthetic Forests, fragile areas) and in any case for vehicle entrance. Restrictions may be implemented during the hunting season in Cyprus, Portugal, Bosnia and Greece. Access on foot is not restricted in any country.

Table 5.1 Existence of legislation to influence public access and other recreational uses

	Bos	Cro	Cy	Gre	It	Por	Srb
Existence of national legislation to influence public access	√		√		√	√	√
Restrictions for public access can be applied in:							
• Private forests	√	√	√		√	√	
• Public forests			√	√	√		
• State forests	√	√	√	√	√		√
Existence of laws affecting recreational uses:							
• Collection of NTFP's	√		√		√	√	√
• Sport activities	√		√	√	√	√	
• Other uses	√		√	√		√	

5.3.2 Forest Functions

The protection of biodiversity, the protection of soil and water, healthy environment and sustainable wood production are some of the forest functions that all countries want to maintain and improve.

The social function of forest recreation is very important for all Mediterranean countries. Firstly, this is because there is an increasing demand for the provision of organized recreational activities which contributes to a better quality of life. It also provides employment for the population of rural areas, since the phenomenon of urbanization is continuous across the region.

Tradition and culture is another aspect which is related to recreation and nature tourism especially in areas which are in or nearby forests. Thus, some countries give great emphasis to sustainable development and focus on special products which are based on a combination of culture and nature, such as rural tourism, nature tourism, cultural tourism etc. For countries which depend economically on tourism it is very important to maintain a good standard of forest recreation, agro-tourism and other associated services.

Generally, in all countries, the management of forests is based on their multifunctional role and their main aim is the provision of non-timber forest products.

5.4 Forest Managers' Education and Tasks Regarding Recreation in Forests

For Cyprus, Italy and Greece planning and management of forest based recreation is considered to be a part of the academic education of foresters. Knowledge of this is improved during their working experience. For the rest of the countries planning and management of forest based recreation is not considered to be part of the basic knowledge and skills of foresters. As is shown in Fig. 5.4, experience plays an important role in forest recreation, planning and design. Also, it is a common situation that all countries lack organized monitoring in the use of recreation facilities and sites as well as monitoring of recreation demand.

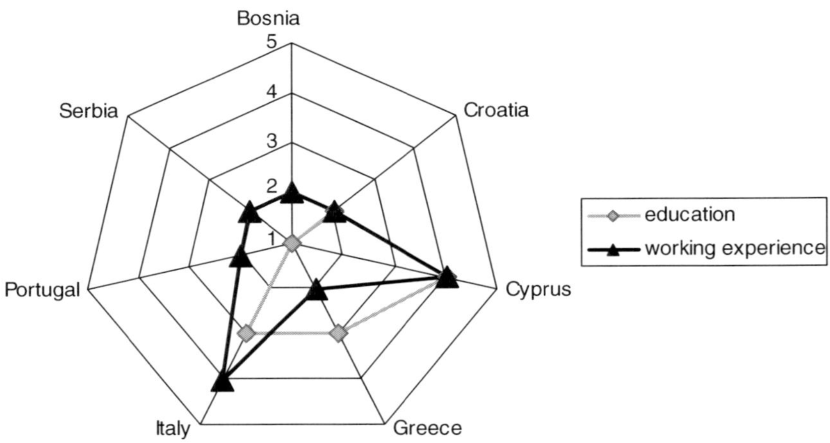

Fig. 5.4: Evaluation of knowledge about planning and management of forest based recreation obtained by education and working experience (1=unsatisfactory, 5=excellent)

Monitoring the use of recreation facilities is carried out in almost all Mediterranean countries in an empirical way. Basically, the authorities involved in looking after the recreation facilities try to satisfy the needs of the users through empirical ways and methods. In most Mediterranean countries the planning, design, establishment and management of forest recreation areas is a major concern of all managers due to the high levels of demand of the public. Most forestry schools (universities, colleges) have now introduced subjects regarding forest recreation, landscape design and many other aspects which have to do with recreation planning, design and management into their teaching syllabus.

5.5 Conflicts

5.5.1 *Types of Conflicts*

As can be seen in Fig. 5.5, the main conflicts in the different Mediterranean countries are between recreation versus hunting/fishing and recreation versus nature conservation. Forestry use (Italy, Serbia, Bosnia and Croatia) and overcrowding effects (Italy, Cyprus) are also important conflicts for some countries. A remarkable point for all countries of the Mediterranean region is that conflicts between different recreational groups are assessed as being very important.

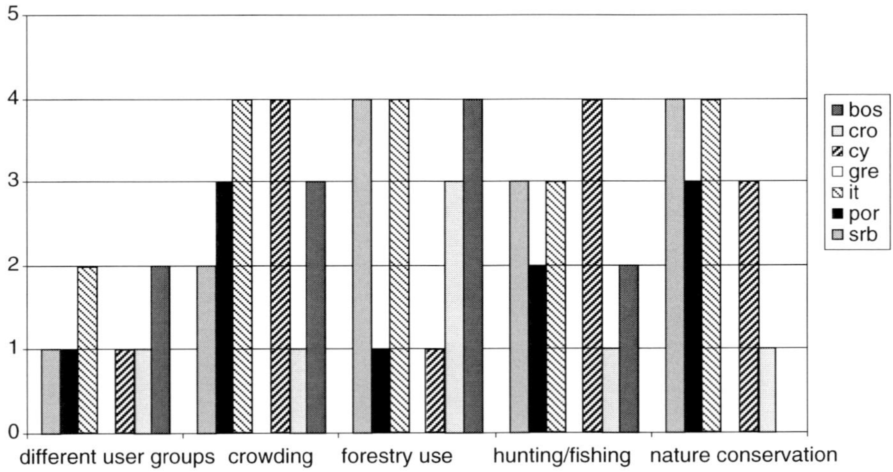

Fig. 5.5 Conflicts levels in the Mediterranean countries (1=not important at all, 5=very important)

5.5.2 *Important Activities and Their Conflicts with Nature Interests*

Due to the climate, the topography and the presence of high mountains in the Mediterranean countries almost all types of recreation activities, including winter sports (alpine skiing being the most popular) take place and at the same time produce conflicts (see Table 5.2 and Fig. 5.6). Popular activities, such as the collection of berries, mushrooms and other non-timber forest products, are also causing conflicts, the main reason being the competition between people. Popular activities that usually take place in summer, particularly during holidays, such as swimming, fishing, picnicking, camping and motorized sports, may cause overcrowding effects and a lot of conflicts between users but also conflicts with the natural environment and nature conservation.

Table 5.2 Important activities and their conflicts with nature interests

Activities	Importance	Countries
Alpine skiing	2	ALL
Ski touring	1.3	BOS, CRO, IT
Cross-country skiing	1	BOS, CRO, IT
Snowshoeing	1	BOS, CRO, IT
Snowmobiling	1.3	BOS, CRO, IT, SRB
Hiking	1.6	Except CY, GR
Nordic walking	1	BOS, CRO, IT
Orienteering	1	BOS, CRO
Geo-caching	1	BOS, CRO, IT
Climbing	1.2	Except CY, POR
Mountain biking	1.2	Except GR
Cycling	1.4	Except GR, POR
Horse riding	1	BOS, CRO, CY, IT
Horse carriage driving	1	BOS, CRO, IT
Motorized sport	1.8	Except CY, GR
Collecting berries etc	2	Except GR
Camp fires / bbq	1.8	Except CY, GR
Picknicking	2	Except GR
Camping	1.7	Except GR
Widllfe / bird watching	1	BOS, CRO, CY
Fishing	2	Except CY, GR
Canoeing	1	BOS, CRO, IT
Swimming	2	BOS, CRO, IT, SRB
Motorized water sports	1.4	Except CY, GR

Figure 5.6 shows that the most important conflicting activities are alpine skiing (in all countries) probably due to the limited space, picnicking, the collection of non timber forest products and campfires.

5.5.2.1 Impacts and Disturbance of Other Societal Activities on Recreation

The main disturbances affecting forest recreation which cause conflicts and are common between the countries of the Mediterranean region (Table 5.3) are: rubbish dumping, visual problems of damaged forests (such as insect damage, forest fire) and the exploitation of natural resources.

Road construction, traffic, industrial and aircraft noise are also important conflicts for some countries. Timber and fuel wood harvesting is not perceived to be a major conflict in some countries because of the general orientation of forest management towards wood production in parallel with a positive trend of non-wood products. In the case of Italy, the conflicts between forest recreation and the maintenance of coppicing and timber harvesting activities are very often a hot issue in forest management and policy and can cause severe arguments between interest groups. Fire breaks mostly affect Greece, Cyprus, Bosnia and Croatia.

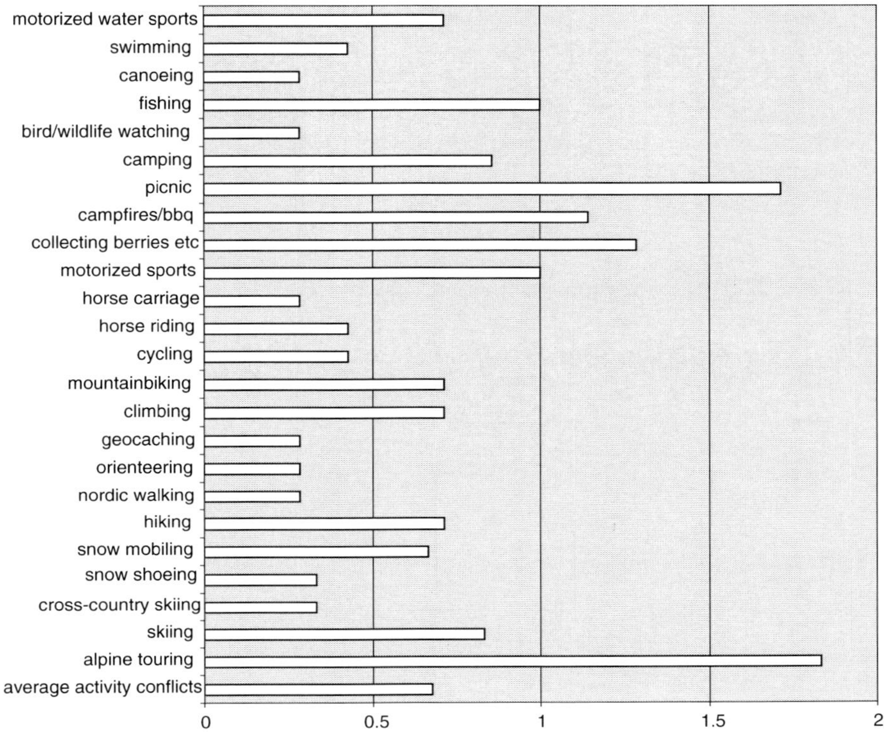

Fig. 5.6 Frequency of conflicts with recreational activities (0=non-existent, 1=rarely, 2=medium, 3=frequently)

Table 5.3 Importance and frequency of disturbance factors on forest based recreation and nature based tourism (importance: 1=low, 3= high, frequency of conflicts: 1= rarely, 3=frequent)

Disturbance	Importance	Frequency of conflicts (only for those countries that observe this conflict)	Countries
Rubbish dumping	2.7	3	All
Visual problems of damaged forests	2.3	2.3	All
Exploitation of natural resources	2.3	2.2	Except GR
Road construction	2	1.5	Except GR
Traffic noise	1.9	1.9	All
Industrial noise	1.8	1.8	POR, CRO, IT, BOS
Aircraft noise	1.8	1.3	CRO, CY, BOS
Large clearcuts	1.8	1.3	POR, CRO, IT, BOS
Fire breaks	1.7	2	All
Other emissions (odur,dust)	1.5	1.2	Except GR
Wind turbines	1.5	1	IT, BOS
Infrastructure for tourism	1.4	1.4	CRO, CY, IT, BOS

Rubbish dumping by private individuals is a "tradition" in almost all Mediterranean countries, leading to a lot of problems. Fire breaks constructed for the protection of the forests from fire is a common practice having adverse aesthetic effects on the forest. Forest fires not only cause visual problems, but in many cases destroy recreation facilities and lead to a need to redesign and re-equip recreation areas.

5.6 Planning, Management and Monitoring

5.6.1 Planning for Recreation and Nature Tourism

In most countries of the Mediterranean region (Croatia, Cyprus, Portugal, Serbia, Bosnia) there is no independent planning for forest recreation and nature tourism (Table 5.4).

Table 5.4 Availability of independent planning (legally required or optional) in the Mediterranean countries

	Bos	Cro	Cy	Gre	It	Por	Srb
Independent planning for recreation and nature tourism in forests	No	No	No	Yes, in special cases	Yes, in special cases	No	No
Legal or optional requirement for forest owners	-	-	-	Optional	Optional	-	-

However, in Cyprus there is a general planning undertaken by the Department of Forests for recreation, particularly during the preparation of development plans for National Forest Parks, which are specific areas where recreation, relaxation, enjoyment and nature study are promoted. Recently, some guidelines were issued by the Department of Forests about recreation under the implementation of the Rural Development Plan. Close cooperation between the Cyprus Tourism Organization, the Forestry Department and other stakeholders has fostered strategic planning and monitoring of forest recreation, rural tourism and nature tourism projects that are in line with the Strategic Tourism Plan which focuses on sustainability.

In the case of Greece and Italy recreational planning is mainly done in special cases. For example in Greece there is a demand, from local organizations and by visitors of the forests found near cities, to provide recreational opportunities. These groups of people exercise political pressure on the government and some times this leads to regional differences in the treatment of the public regarding forest based recreation and nature tourism.

5.6.2 Planning Steps

Although there is a long tradition in using the forest for recreation, particularly during summer, planning based on social and ecological aspects is something that is ignored by most Mediterranean countries, except Greece and Italy. From Table 5.5 it can be seen that most countries understand planning to be only what the planners and managers think and it is based on empirical methods rather than modern scientific approaches.

Table 5.5 Steps that are undertaken for planning forest recreation at regular intervals in the Mediterranean countries

Steps in planning forest recreation	Countries
Inventory of ecological conditions	GRE, IT, SRB
Inventory of social conditions	GRE, IT, SRB
Analysis of conflicts	GRE, IT
Definition of vision and goals	GRE, IT
Participation in the definition of goals	GRE, IT
Definition of standards	-
Deduction of measures	GRE, IT
Participation in the deduction of measures	GRE, IT
Monitoring	-

There is also some public participation in the definition of visions and goals by actors such as land owners, local residents, interest groups and organizations, community representatives etc. Monitoring of such areas is carried out at irregular intervals and there are guidelines issued by some forestry services which direct the planning and management of forest recreation.

No country has developed any tool or method of monitoring recreation demand or the expectations of visitors of forest recreation. In Greece, Italy and Serbia, monitoring is carried out at irregular intervals (whenever necessary). In Cyprus monitoring is at the very first stage of development.

It seems that in all countries the supply of recreation facilities is reactive to local demand or opportunities seen by local staff and is based on the experiences of the local officers. Forest recreation planning in private forests seems to be of no concern in any Mediterranean country.

5.6.3 Overall Evaluation of Recreation and Tourism in Planning and Monitoring

Each country was asked to evaluate the integration of forest-based recreation and nature tourism in the planning and monitoring of the country's forests. From this current state of the Mediterranean countries is assessed as being below average. This is a result of several weaknesses in the majority of the countries, leading to weaknesses in fulfilling the demand for good quality recreation opportunities and to the inadequate planning, monitoring, legal support and education of management personnel.

5.7 Recreational Infrastructure

5.7.1 Supply of Basic and Additional Infrastructure

In urban and rural areas the supply of basic infrastructure is assessed as either/both good or very good in all the Mediterranean countries (Fig. 5.7a). On the other hand, the supply in remote areas is considered to be satisfactory (good) only in Cyprus, Portugal and Italy. The provision of access is generally good in all countries except Croatia where is somewhat lower than satisfactory. It is surprising that countries with a long tradition and a high level of demand for forest recreation have not developed proper mechanisms for the supply of reasonably good infrastructure for recreation.

Greece, Cyprus and Italy evaluate the supply of additional infrastructure in urban and rural areas as good or very good (Fig. 5.7b). This is mostly because the attractive locations (high mountains, coasts and dense forests), in combination with hot and dry summers lead local people to prefer such areas for recreation. Also, the large number of local and foreign tourists, especially during the summertime, increases

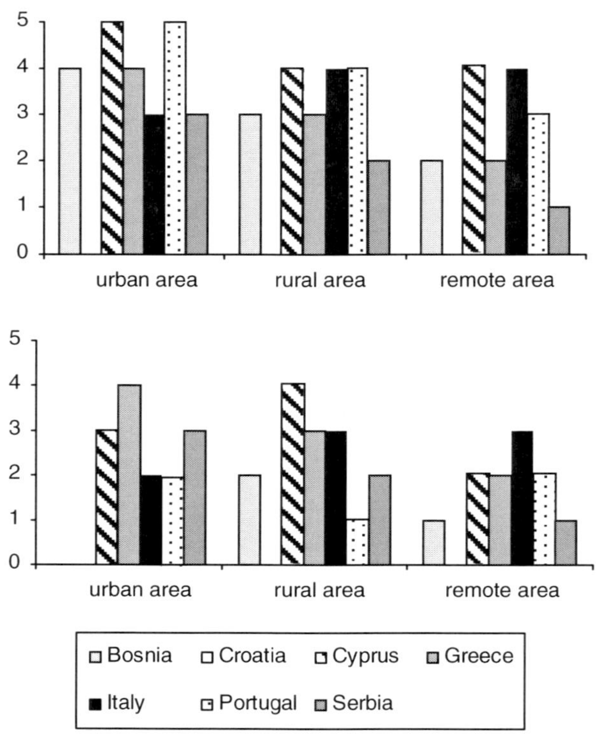

Fig. 5.7 Supply of basic infrastructure (*left figure*, **a**) and additional infrastructure (*right figure*, **b**) (no bar = non-existent, 1=unsatisfactory, 5=very good)

the pressure on these sites (leading to overcrowding), so that forest owners and managers are forced to supply more facilities in these areas in order to meet the demand and to protect vulnerable areas. Croatia assessed the supply of additional infrastructure in urban areas as also being good but at the same time being poor in rural areas. Portugal reports the supply of additional infrastructure as being poor in urban and unsatisfactory in rural areas. The explanation for this is that the majority of forest resources (83%) are in private ownership.

The supply of additional infrastructure in remote areas is assessed as being poor to unsatisfactory for the majority of the countries. In the case of Italy, the fairly good state of additional infrastructure in remote areas is very linked to the longstanding tradition of nature tourism (trekking and climbing) as well as skiing in quite remote mountain regions

5.7.2 Additional Facilities

From Fig. 5.8 it appears that there is a significant gap between remote and urban areas in terms of the additional facilities. However, the gap between urban and rural

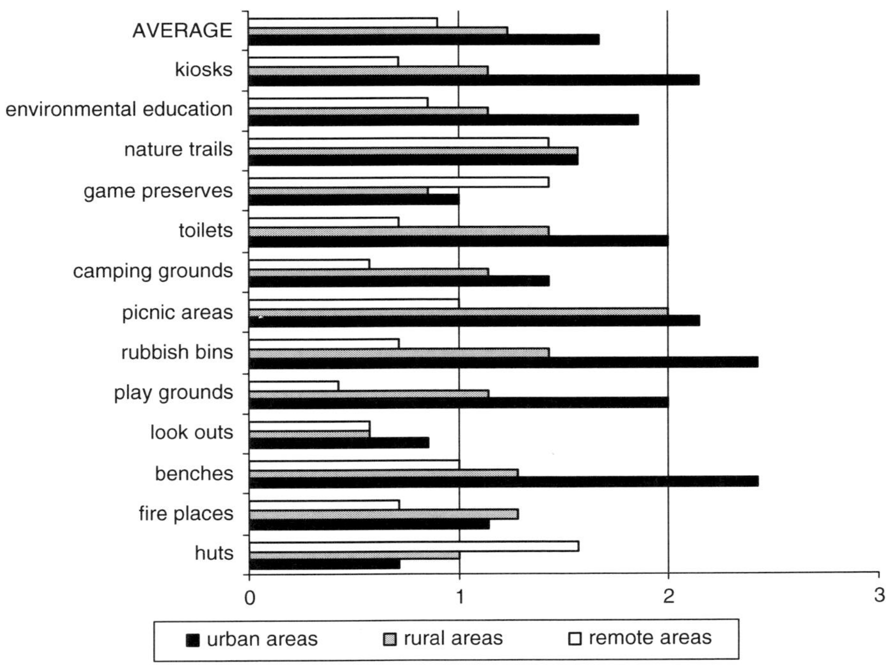

Fig. 5.8 Occurrence of additional infrastructure in urban, rural and remote areas in the Mediterranean countries (0=non-existent, 1= rarely, 2=reasonable, 3=frequently)

areas is not as great and almost all Mediterranean countries emphasise the supply of satisfactory number and type of facilities in the areas used for recreation. Kiosks, toilets, picnic areas, rubbish bins, play grounds and benches are all facilities which are supplied in quite large numbers and it seems that they generally meet the requirements of the public.

5.7.3 Quality of Infrastructure

Good quality equates to good original design, regular renovations, maintenance, checking of safety features and adaptation of designs over the years.

The quality of infrastructure particularly in the rural and remote areas is unsatisfactory in all Mediterranean countries, except Cyprus (Fig. 5.9). Some countries such as Croatia, Cyprus and Greece, keep a very good standard in urban areas, where people from the cities visit and use these recreation facilities. It also appears that in most countries there is a lack of resources for the proper maintenance of recreation facilities. Likewise, the maintenance, the quality and the proper design of recreation facilities is of low priority to many forest managers.

In the case of Cyprus, the infrastructure for forest recreation is funded, managed and maintained mainly by the Forestry Department. This fact leads to good results in terms of the quality of infrastructure and makes management easier. This is mainly explained by the fact that Cyprus is very small in size, when compared to the other countries. In the case of the other Mediterranean countries, the lack of a central service to evaluate and look after the quality of recreation infrastructure is the main reason for the unsatisfactory conditions of the facilities.

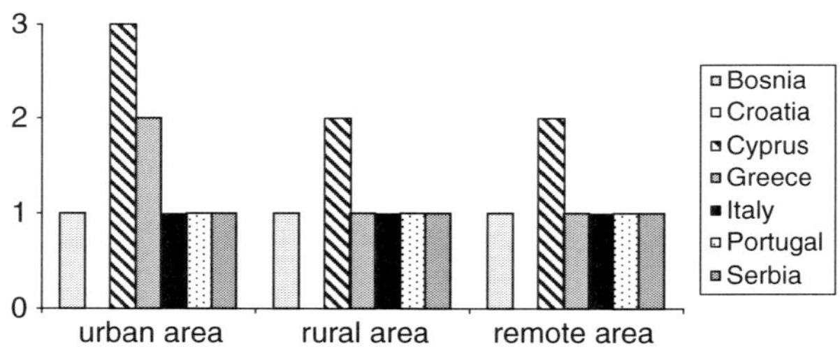

Fig. 5.9 Quality of infrastructure for recreation and nature-based tourism (no bar = non-existent, 1=unsatisfactory, 3=excellent)

5.8 Approaches to Visitor Management

The most common actions for the management of visitors are: restrictive management actions, soft management actions and facilitating management actions. The following figures and tables show the frequency and occurrence of use of different types of management actions in the Mediterranean countries in urban, rural and remote areas and the level of acceptance of these actions by the public.

Management actions are intended to protect and maintain recreational infrastructure and services in order to be attractive to the public and to protect the environment from risks as a result of human mistakes and carelessness. They are also intended to help people towards an understanding of nature and its importance in their life. The various management actions are classified into intensive management actions (regulations, prohibitions), soft management actions (infrastructure, signposting) and guiding and educational management actions (education – training, guiding) (see Fig. (see Fig. 5.10 and Table 5.6).

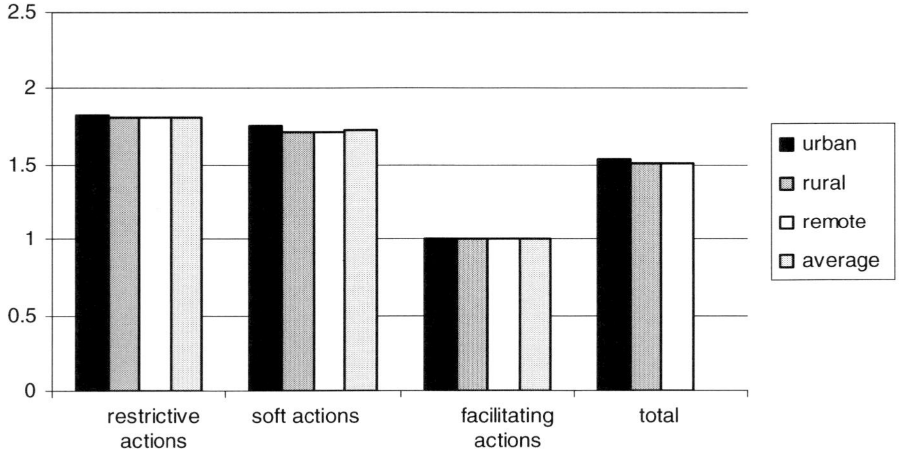

Fig. 5.10 Occurrence of management actions in the Mediterranean region, in general and in urban, rural and remote areas (0=non-existent, 1= rarely, 2=regularly, 3=frequently)

5.8.1 Restrictive Management Actions

It appears that all the countries use intensive management actions as tools to protect nature and the recreation facilities from improper use by visitors. Although these actions are generally of a prohibitive nature, they have a relatively high degree of acceptance by the public (Table 5.7). This shows a degree of maturity on the part of the public and also that the management (administration) is generally doing a good job in the provision of publicity and information. The most frequent intensive management actions are rules and regulations, prohibitions and the marking of protected areas and taboo zones. Certain rules and regulations (such as the lighting of fires) and prohibitions (such as restricting entrance to fragile areas or hunting in certain areas) are not popular in all Mediterranean countries.

Table 5.6 Occurrence of management actions (1=rarely, 3=frequently)

		Occurence of management actions		
Management actions	Countries	Urban areas	Rural areas	Remote areas
Restrictive management actions				
Rules and regulations	All	2.6	2	1.7
Prohibitions	All	2.3	2.1	1.9
Entrance fee	All	1.7	1.8	1.7
Marking protected areas and tabu zones	All except Gr	2	2	2.2
Barriers and fences	All	1.7	1.9	1.3
Limitiing visitors entrance	All	1.5	1.6	1.4
Soft management actions				
New attractive infrastructure	All	2	2	1.6
New infrastructure for environmental education	All except Gr	1.7	1.7	1.5
Signposting for visitors to act in responsible way	All	1.9	1.7	1.7
Ranger systems	All except Gr, Por	1	1.2	1
Facilitating management actions				
Guiding visitors by thematic tours	All except Gr, Cy	1.3	1.3	1.5
Providing special training for user groups	All except Gr, Cy	1.8	1.5	1.8
Providing special training for commercial providers	All except Gr, Cy	1.4	1.6	1.6

Table 5.7 The level of acceptance in the Mediterranean region for the various management actions (high >= 2.5, medium > 1.5 and < 2.5, low =<1.5)

Acceptance by the public	Restrictive management actions	Soft management actions	Facilitating management actions
High		Attractive infrastructure Educational infrastructure	
Medium	Rules and regulations Prohibitions Entrance fee Marking protected areas/tabuzones Barriers and fences Limiting visitors' entrance	Signposting responsible visitor behaviour	Guiding thematic tours Guiding ranger systems Training special user groups Training commercial providers
Low			

5.8.2 Soft Management Actions

The installation of new attractive infrastructure, the proper signposting and the infrastructure for environmental education are management tools which are used to a high degree by all countries (see Fig. 5.10). People are very willing to accept the new infrastructure as well as the proper signposting in urban and rural areas, but they are not so demanding of them in remote areas. The use of a ranger system is also accepted well by people but it is very limited in its use, the explanation for that being the high cost of employing rangers to guide people and to look after recreation sites and structures.

5.8.3 Facilitating Management Actions

These management actions seem to have low or medium importance for the managers (see Fig. 5.10). The main reason is probably the extra efforts and additional personnel needed to implement such actions.

5.9 Future Trends and Challenges

5.9.1 Trends and Challenges

The main trends and challenges for the Mediterranean region can be summarized as follows:

- There is an increasing demand for forest recreation from the local population as well as from the many tourists visiting these countries.
- In order to succeed in the competitive world tourism arena, most countries are developing sustainable and unique products by capitalizing on the utilization and promotion of and at the same time preservation of their forests and other nature assets.
- The trend is for nature tourism to progressively increase over this decade, as people are becoming more environmentally conscious and wish to explore the unique natural attractions of various destinations.
- Environmental education seems to be given inadequate emphasis by the competent authorities and this creates problems in the information given to the public on the one had and the proper education of the responsible persons for the design and implementation of forest recreation on the other hand.
- The improvement and upgrading of recreational infrastructure is one of the weaknesses of almost all countries. There is a changing demand in all Mediterranean countries regarding both the quality of recreation facilities and the variety of recreation opportunities. Policy makers should give much more attention to this and consider making more funds and personnel available for this.
- The monitoring of and research into recreational demand is almost non existent. More emphasis on this is needed in order to improve the recreation standards and to satisfy the recreation demands.

5.9.2 Specialities in the Region

Due to the great number of tourists visiting recreation opportunities and facilities, the climatic conditions prevailing in the region and some other characteristics, there is a number of specialities which make forest recreation in the Mediterranean countries more vulnerable:

- Forest fire is a common phenomenon during the summer when recreation demand is at its peak. This threat endangers the existence of forest areas devoted to recreation, causes damage to or destruction of recreation infrastructure and endangers the lives of visitors.
- The impact of recreation on the landscape is quite high and many areas suffer locally as a result of visitor pressure. Pollution through littering, for example, is also a problem which needs a lot of effort to solve.
- Forest recreation and hunting are two popular activities, particularly in the mountainous areas in the region. These two activities are not compatible in certain cases. It appears that policies and legislation still need to be improved in order to cope with this problem.
- The design for forest recreation is basically in the hands of the governments through the forest services of each country. The local authorities deal mainly with the provision of local parks. The resulting lack of coordination or cooperation results, in many cases, in poor recreation opportunities and facilities in particular locations.
- In Greece overgrazing creates special problems for forest recreation.

Chapter 6
Eastern Region

Agata Cieszewska, Dóra Drexler, Péter Kalincsák, Olgirda Belova, and Jan Prochazka

6.1 Introduction

The Eastern European countries presented in following report consist of six countries distinguished mostly by their geographical location. The Czech Republic, Hungary, Latvia, Lithuania, Poland and Slovakia reflect contemporary political and administrative conditions and all lie along the eastern border of the European Union. They are related not only by their setting, but also historical factors. Estonia joined the Cost Action towards the end so there is no data available (Fig. 6.1).

Fig. 6.1 Countries within the Eastern region

A. Cieszewska (✉)
Warsaw University of Life Sciences (SGGW) Department of Landscape Architecture Ul. Nowoursynowska 159, 02- 786 Warsaw, Poland
e-mail: agata_cieszewska@sggw.p1

U. Pröbstl et al. (eds.), *Management of Recreation and Nature Based Tourism in European Forests*, DOI 10.1007/978-3-642-03145-8_6,

Geographical aspects of Eastern Europe determine the prevalence of mixed forest in the northern part, similar to those of the Nordic region and also in the southern part in the Carpathian mountains, where spruce was introduced even in lower mountain locations. In the central part of the region Scots pine, spruce, beech and oak are the dominant forest species.

In all the Eastern countries there are similar influences on forest recreation conditions. The first aspect is that after the Second World War and for the next 45 years, most forest belonged to the state, and public access had been free, except for restrictions in large military zones. This meant that recreation was not part of the revenu generating aspects of forestry and it also meant that people viewed state property as something they could use as they wanted.

Traditionally forest recreation in Eastern Europe has a strong relationship with the collection of the fruits of the forest floor, especially berries and mushrooms; for people in some countries it also provides and important source of income. Even during the Communist system these aspects of forest recreation had been popularized by excursions organised by companies for their workers. Recent years have brought a more individual approach to forest recreation. The increasing use of private cars has made this access easier, but also caused some new threats. Forest recreation activities have some regional characteristic and traditions, with cross country skiing and down hill skiing being much more popular in the Czech Republic, Slovakia and the Baltic Countries, than in Hungary or Poland.

6.2 Forest Cover, Development and Ownership

6.2.1 Forest Cover

Forest is the climax ecosystem in all of the Eastern countries of Europe. In prehistoric times the forest area would have covered almost 90% of the land area. Over the last few centuries the expansion of agriculture and other human impacts have reduced the forest area as it has everywhere else. Deforestation stopped in the early twentieth century.

In the Eastern countries the average proportion of forest cover is about 33%, so it is close to the European average. The proportions of forest cover differ from country to country, but not widely (Fig. 6.2).

The location of most forest cover is mainly related to geographical conditions. The climate and in some cases the soil conditions of northern locations (Latvia) and higher altitude areas (Slovakia, Czech Republic, southern Poland) are disadvantageous for agriculture and therefore forest tended to remain. The least forested among Eastern European countries is Hungary where natural conditions are most suitable for agricultural activities – mild climate, wide plains, and fertile soils. The Great Hungarian Plain is less forested than other parts of the country, in consequence of which the forests are not spread evenly here and only 2/3 of the population can

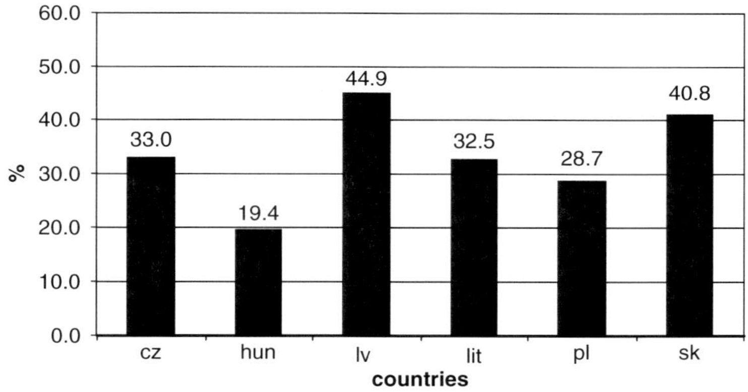

Fig. 6.2 Forest cover (%) in the Eastern countries

reach forests easily for recreational purposes. Similarly in Poland, where natural conditions favour agriculture and the vast plains in the central part are the most deforested. As well as natural influences, the effect of historical factors is significant in patterns of forest dispersion. Forests in Eastern European countries are often fragmented, particularly in Poland where more than 6,000 forest patches each cover less than 5 ha in area (Fronczak 2003). In other countries woodlands are distributed more evenly.

Forest area per capita is an important statistic to help evaluate the value of the countries' forests. The Latvian proportion is markedly different from other Eastern European countries – 1.23 ha/per capita. Latvia and Estonia (and to a lesser extent Lithuania) are more like Nordic countries, but only the Latvian proportion is close to the European average at 1.4 ha/capita (State of Europe's forest 2003), and it is several times greater than the proportion of other countries of the group (0.18 ha/capita in Hungary to 0.5 ha/capita in Lithuania), while the EU average is 0.30 (Distinguishing the Wood 2004).

6.2.2 Development

The forest area in the East European countries is expanding in line with the overall European trend. Expansion of forest cover is significant different in Poland, where it is increasing at the fastest rate of this group of countries. This situation arises from the spontaneous recolonisation of forest on poor quality soils – mostly abandoned land – of no future value for agriculture use. In Latvia and Estonia abandoned lands are also reverting to forest quite quickly. Reforestation in the mountain regions – for example the Carpathians – is connected to overgrowing of mountain pastures. The lowest rate of expansion is taking place in the Czech Republic and Slovakia, but forest cover in those countries, beside Lithuania, is amongst the highest in the region after Latvia and Estonia. After 2004, and the accession to the EU, in some countries, for example Poland, the opposite trend can now be observed, as former arable fields reforested during the last 10 years are being reconverted back to agricultural use, because of the additional EU farming support.

6.2.3 Ownership

The ownership patterns of forests have changed radically over the last 15 years. Private forests were almost unknown between 1945 and 1990. After the Second World War private forest estates were nationalized in most countries, only in Poland did the process affect mainly the large and medium size forests with the smallest forest patches remaining in private hands. Most of Eastern European countries implemented the land reform after 1990 and forests have been returned to their former owners or their inheritors (Fig. 6.3) (Marghescu and Anderson 1997).

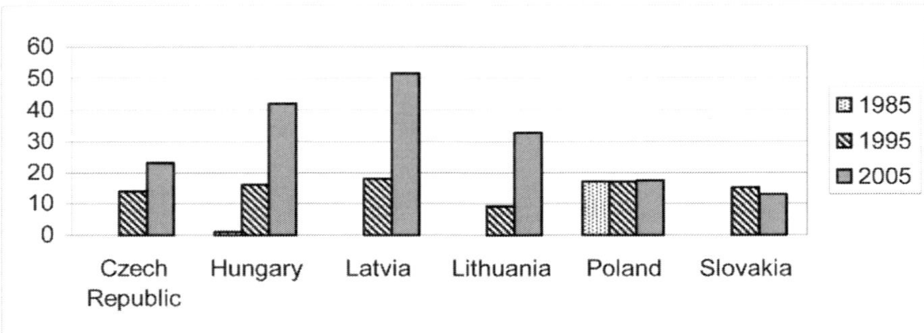

Fig. 6.3 Comparison of private ownership of forest land in Eastern European countries before (1985) and after (1995, 2005) political transformation (%) (source: Marghescu and Anderson, 1997)

Today in the Eastern European countries the state sector (and some community forests) dominates except in Latvia and the Czech Republic where there are more private forests. This is not a stable situation, with many private owners still to have their forests return – for example 17.5% of forests are reserved for restitution in Lithuania. Latvia has no separate data of community owned forest. After the transformation of the political systems the discussion about the vision of forest ownership remains unfinished. This debate continues depending on the fluctuating political situation – for example during the elections in 2005 and 2007 in Poland.

The proportion of state-owned forests is very significant in all countries, but there are differences (Fig. 6.4). The value is varying from more than 80% in Poland to about 40% in Slovakia. Over the last 15 years the major change of state forest cover has concerned mostly Slovakia (from 92 to 43%) and the Czech Republic (from 96 to 60%) and less so in Lithuania, Latvia and Hungary (Ferenc 1996). The community owned forests are irrelevant (ca.1%) in Hungary and Poland, but almost

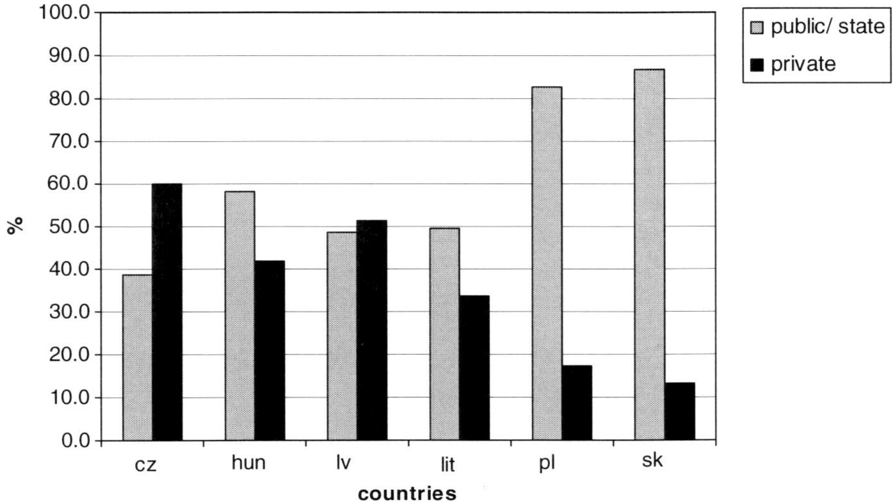

Fig. 6.4 Forest ownership in the Eastern countries in 2005 (%)

half of the forests are owned by communities in Slovakia (the proportions shown in Fig. 6.4 hide the large amount of community forest within the figure for public/state forest). The ratio of owners is still changing in Baltic countries.

Size categories of private forests in Eastern European countries are different. In Lithuania and Poland all private forests are below 5 ha in area. In Hungary in 35% of forests are below 5 ha and in the Czech Republic 27%. In Slovakia there is no such very small ownership pattern (less then 0.1%). The larger private forest estates with over 100 ha in area account for 61%, in Slovakia, 32% in the Czech Republic, and only 3.5% in Hungary and there is no such category in other countries (source: Marghescu and Anderson 1997). These circumstances reflected different ideas in the restitution and/or privatization processes of each country. The structure of private forests with a domination of small ownerships creates problems for forest management as well as recreational access.

There are no regional differences with regards to ownership in Czech Republic, Hungary and Slovakia. In the others (Latvia and Lithuania) there are but only in few areas. In Poland regional variations in ownership patterns reflect the border changes that took place after the Second World War. Northern and western parts of Poland are dominated by state forests. In central and eastern Poland as well as in the mountains (Carpathians) the private forest proportions range from 10 to 50%. The size of private forest holdings does not vary widely in the Czech Republic and Slovakia. There is some regional diversity in other parts of the Eastern Europe.

6.3 Legal Background, Public Access and Forest Functions

6.3.1 Legal Background

Recreational use of forests is affected by law in all of the Eastern European countries. In Hungary, Latvia and Poland regulations contain general directions regarding forest recreational use, while in Slovakia and in the Czech Republic there are several acts pertaining to recreational uses. Lithuania has both general and specific directions in its legislation.

All six countries have in common the fact that they do not have regional variations of forestry legislation. This can be explained by the different historical background of the countries compared to the Central and Atlantic countries. The Eastern European countries did not have a federal structure in pre-communist times. They developed as nations with centralized governance. This may change in the direction of a stronger regionalism within the European Union.

6.3.2 Public Access

The legislation of all countries covers aspects of public access, and the collection of mushrooms, berries and minerals. Sport activities (e.g. horse riding or mountain biking) are included in Hungarian, Polish, Slovakian, Czech and Lithuanian legislation. Further aspects covered by legislation are hunting in Hungary, Lithuania, the Czech Republic and Slovakia. Nature protection, ant-hill damage, dog walking, noise, grazing and rubbish are mentioned by Poland as being covered by legislation while building activities as well as the utilization of water sources are emphasized in Slovakia. In Lithuania all activities mentioned above may be restricted in certain areas of forest according to the Lithuanian Law of Protected Areas. In Hungary there also exists the possibility of restricting activities in parts of forests because of measures of nature and/or forest protection.

The following Table 6.1 gives an overview of the legislation having an influence on public access and other recreational uses.

Table 6.1 Existence of legislation having an influence on public access and other recreational uses

	Cz	Hun	Lv	Lit	Pl	Sk
Existence of national legislation to influence public access (yes/no)	Yes	Yes	Yes	Yes	Yes	Yes
Restrictions for public access can be applied in:						
• Private forests	No	Yes	Yes	Yes	Yes	Yes
• Public forests	Yes	Yes	Yes	Yes	Yes	Yes
• State forests	Yes	Yes	Yes	Yes	Yes	Yes
Existence of laws affecting recreational uses:						
Sport activities	Yes	Yes	Yes	Yes	Yes	Yes
Collection of NTFP's	Yes	Yes	Yes	Yes	Yes	Yes
Other uses	Yes	Yes	Yes	Yes	Yes	Yes

From the above data, the conclusion can be drawn that forest recreation is addressed in the legislation of all six countries with regard to the different thematic aspects within forest recreation. However, mostly it is general directions that are given in regulations (except for Slovakia and the Czech Republic) and the range of aspects taken into consideration in legislation is not broad enough (e.g. in Latvia). The fact that legislation has a nation-wide scope in all six of the countries probably makes the development of the legal regulations easier for the future.

6.3.3 Forest Functions

All of the Eastern countries have a functional classification of forests in their legislation. These classifications are considered to be very important for the spatial planning process in all countries and accordingly have to be taken into account. However, in reality some negative examples also exist where the strict legislative classification of forest functions is ignored.

The categories of protected areas offered by the questionnaire, that is protected areas for landscape conservation, nature parks and the national parks, all exist in the eastern countries, except for Lithuania, where the category of Nature Park is not present. Furthermore, smaller-scale protected areas are also mentioned by each country. These are: natural monuments in Hungary and in Slovakia (plus national natural monuments), micro-reserves for protection of biotopes and species in Latvia and Slovakia, nature preserves in Poland, in the Czech Republic and in Slovakia (plus national natural preserves). Slovakia has in addition a category of protected sites. Lithuania seems to have a quite unique categorization of protected areas among the Eastern countries, because the following types of protected areas are listed: strict nature reserves, biosphere reserves, biosphere grounds, state nature reserves for biodiversity conservation, state reserves of other purposes (geologic, geomorphologic, landscape purposes), regional parks with designated and managed zones for education, recreation, nature conservation, protected objects of nature heritage and protected areas with the priority of nature reconstruction and ecological protection. The last category consists of forests with restricted forestry activities. Latvia also has biosphere reserves. All countries have (or will have to designate as soon as possible) Natura 2000 areas. The designation of Natura 2000 areas is relevant for all EU member states, so includes all of the eastern countries.

Four main forest functions: timber production, protection (soil, water and avalanches), recreation and biodiversity protection have received different emphasised in Eastern European countries. In the Czech Republic, Slovakia, Lithuania and Latvia timber production is the main forest function. Economic use predominates also in Hungary's forests (65%); a legal category of forests with the priority of health, social, tourism, education and research connected functions encompasses only 2% in the forests in Hungary, although, 20% of forests are protected. They play the significant role in forest recreation in Hungary. In Poland the multifunctional model of forest function is promoted in national forest policy: ecological, economic and social, but 53.2% of forest cover has the economic function (timber production) as the primary objective; 46.8% of the forest is protected (soil, water, other) but there is no particular category for the recreational function alone. In general, in all countries, the role of recreation in forests is higher near the bigger cities. Lithuania emphasises the recreational potential of forests, particularly for nature tourism, in the national and regional parks as well as in the forest enterprises.

6.4 Forest Managers' Education and Tasks Regarding Recreation in Forests

The education focused on planning and management of recreation in forests is the part of the basic knowledge of foresters in the Czech Republic, Lithuania, Slovakia and Poland. This corresponds with the traditionally high level of forester's education in the past. Latvia is excluded from this group with an absence of formal education in this sense (as is Estonia). The following figure shows the evaluation of forest managers' knowledge about planning and management of forest based recreation.

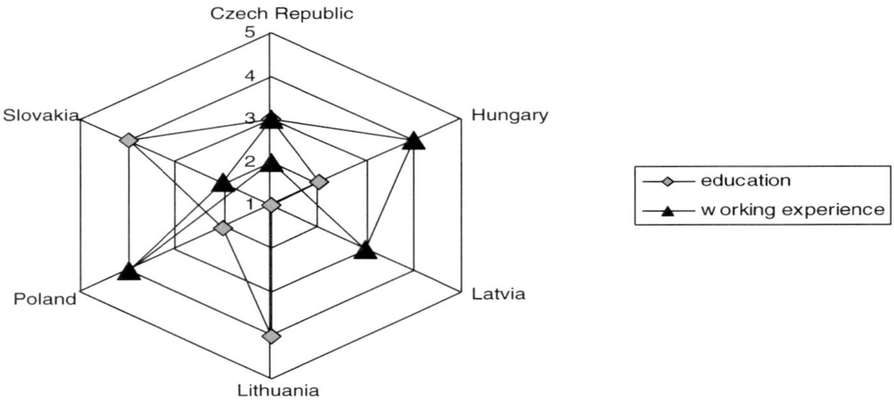

Fig. 6.5 Evaluation of knowledge about planning and management of forest based recreation obtained by education and working experience (1=unsatisfactory, 5=excellent)

The significance of education and implementation of recreational concerns in forest management and planning is connected to the number and quality of high schools/universities in each country. In most countries – Hungary, Latvia and Poland – working experience is considered to be more important than education. The Czech Republic has a similar rating for both, and only Slovakia considers education to be better than working experience in recreational management. Lithuania also considers that there is quite a good situation in education.

The tasks of the forestry administration are the same in all countries – mainly planning and the design of monitoring frameworks, with exception of Lithuania which has an additional task of forestry inventory.

All countries consider the main tasks of management to be planning, information, environmental education, management actions for visitor guidance, supervising and establishing infrastructure. This preference for integrated tasks seems quite a strong aspect of recreation management. In Lithuania the precise tasks of forestry management depend on the functional purposes of the particular forest type according to the Forest Law and the corresponding forest management plan.

6.5 Conflicts

The recreational function of the forests has many aspects (Supuka and Vreštiak 1984):

- relaxation – based on escaping from reality, seeking silence, fresh air;
- sports – when the forest is used for summer as well as winter sports such as running, biking, cross country skiing, down hill skiing;
- obtaining material products as from hunting, fishing as well as mushroom and berry picking;
- cultural – discovering cultural values of the natural environment, participation in artistic activities – taking photographs, painting;
- science and nature study (flora, fauna, ecosystems).

All these aspects of use by human beings influence nature while nature influences humans. This process acts in the short term as well as over the long term.

6.5.1 Types of Conflicts

There are five categories of different types of conflicts concerning recreation:

- conflicts between different recreational groups, for example when the same area or trail is multifunctional, can arise between hikers and horse-riders and/or bikers. However, not all multifunctional trails bring confrontation – some of them are used in different periods of the day or seasons. This problem also includes the conflict between relaxation and active sports;
- conflicts due to overcrowding – when the capacity of the trail is exceeded – the problem can arise periodically or be quite local in its effect;
- conflicts between different land uses, usually between timber production and recreation, when safety requirements can create a conflict with the aesthetic values of the forest. This kind of conflicts is usually temporary, but in some locations can be permanent, for example when reforestation blocks long distance views;
- conflicts between recreation and other uses in the forest, e.g. hunting and fishing – these activities can be considered as part of the economic aspect – but their result are not permanent;
- conflicts between recreation and nature conservation that may be only occur locally within protected areas or concerning protected objects (plants, animals, others).

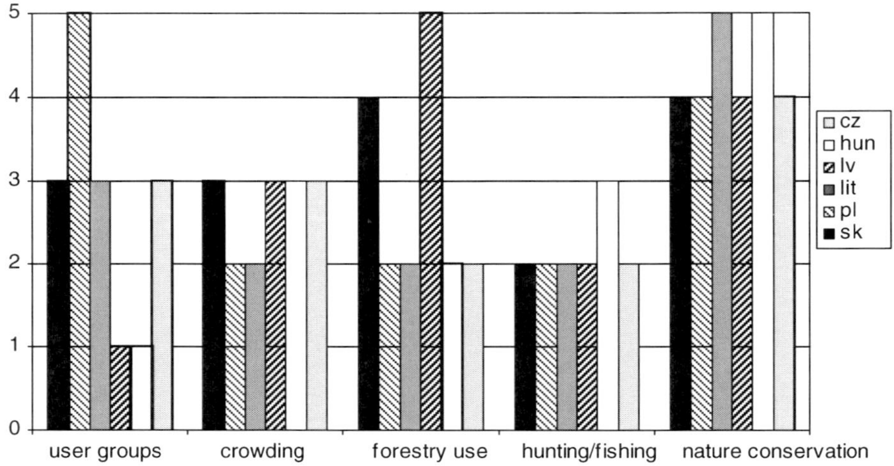

Fig. 6.6 Conflicts levels in the Eastern countries (1=not important at all, 5=very important)

In the Eastern European countries the most common conflict is between recreation and nature conservation (Fig. 6.6). Most protected areas were established in Eastern Europe during recent decades (except for Slovakia, where the first national park was established in 1948), many of them in traditional recreational areas, e.g. in Latvia along the Baltic coast; also in this group of countries perceptions of forests are still strongly connected with customary collecting of mushrooms and berries. The feature of new protected areas is a lack of infrastructure, therefore the combination of access restrictions with tourist attractions initiates conflicts. In each country this problem has a different aspect. In Hungary it mainly affects national parks while in Poland and Slovakia the problem concerns the change in use of formerly easily accessible woodlands for cycling, winter skiing, riding horses but also collecting mushrooms and berries.

The next problem but generally considered to be of medium importance by almost all countries in this group is the problem of overcrowding. Overcrowding mainly affects areas close to the bigger cities in Poland and Slovakia, in city forest parks and main resorts in Lithuania and Hungary or in the centres of winter sports in the Czech Republic. Latvia finds this problem in some very popular specific areas among special tourist forest trails that result in substantial damage to nature values and in the loss of aspects of their uniqueness.

Conflicts between recreation and forestry activities has a different meaning in Eastern European countries, but the activities listed as causing conflicts shows that they not new problems. For four countries recreation versus forestry use has low level of conflict. In Lithuania and Poland it arises from multifunctional forest usage so that timber production is seen as being opposed to collecting mushrooms and berries, but in the Czech Republic and Hungary the most significant problem is littering in forests. This problem also has a high importance in Latvia not only because the population is losing traditional mushroom picking places, but also because of serious landscape changes due to intensive timber harvest near summer house areas located in or close to the forest and along tourist trails. In addition Slovakia identifies this problem as being significant mostly as a result of illegal fires, which are at present strictly prohibited. In addition, in some parts of the region income from berry collecting is higher than the forest stand value, e.g. Poland and Latvia.

The problem pointed out as crucial in many European countries between recreation and different recreational groups is almost irrelevant in Eastern Europe. Participants of this conflict are indicated the same in all six countries with examples of activities as cycling/walking, motorcycling/walking, winter hiking/cross-country skiing. The problems are concentrated around trails used by too many different activities. Of less importance are conflicts between recreation and hunting/fishing, just Hungary classifying it as medium because of disturbance to game.

6.5.2 Important Activities and Their Conflicts with Nature Interests

Some types of recreational activity are highly associated with particular natural features. Landscape differentiation but also national/regional traditions in most cases reflect different degrees of conflict between recreation and the natural environment and nature conservation. The next table shows the frequency of conflicts for different recreational activities.

The most significant conflicts in this group of countries concern two activities. The first is motorized sports, that are officially only permissive in the forest on public roads, but it is extremely popular in Eastern European countries, especially in Hungary, Lithuania and Latvia. The conflicts, which are caused by these activities, are high. Latvia reported it as illegal access to restricted sites, especially 4×4 use in sand dune areas, and it also has an important role in Hungary. Motorized sports are becoming more and more popular as off road trips using 4×4 vehicle and quad bikes; therefore this low acceptance of regulations within the forest is bringing new types of recreation conflict to Eastern European countries.

The second important conflict with other recreational activities is reported to be picnicking. The problem is a consequence of relatively low level of infrastructure, thus people use also unprepared areas where the managers may not want picnicking to take place, but there is a different type of conflict occurring around big cities where real estate development takes place on public space traditionally used as picnic areas, thus reducing the opportunities available to people.

The highest frequency as well as severity of conflict in Eastern European countries involves collecting berries, mushrooms, minerals and enjoying camp fires. There are some regional differences. The most frequent collecting activities are in

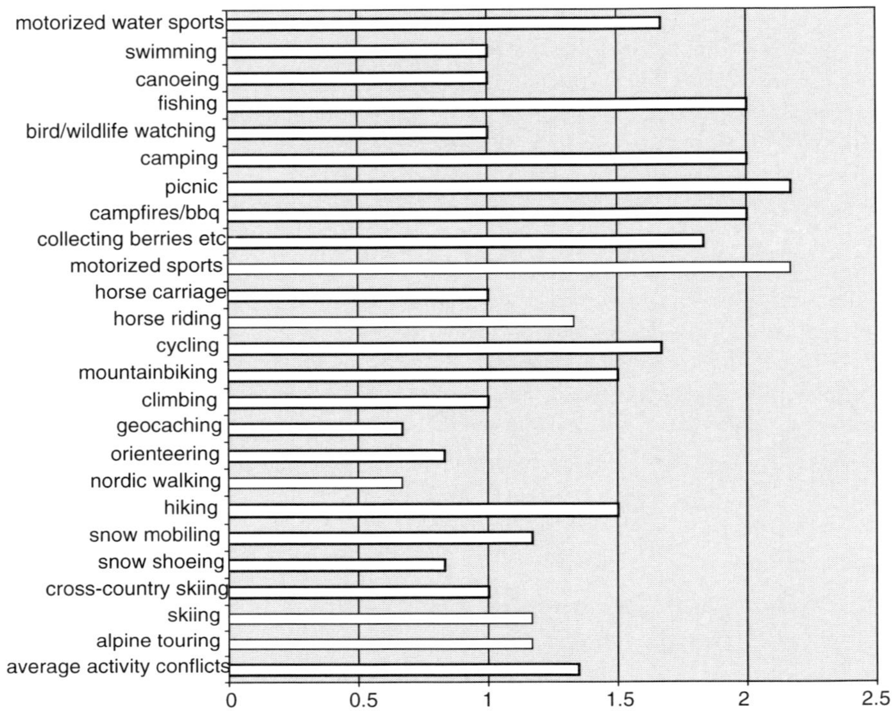

Fig. 6.7 Frequency of conflicts with recreational activities in the Eastern countries (0=non-existent, 1=rarely, 2=medium, 3=frequently)

Lithuania, Poland and Slovakia, where collection of berries and mushrooms has long tradition. People are aware of some damage to the forest caused by recreation, but they do not necessarily connect it with collecting berries and mushrooms (Tracz and Gaca 2000). Even a lack of rare mushroom species is not enough of a warning sign to limit this huge movement in summer and autumn, when whole families come into the forest for mushroom picking. Nature protection measures are not strong enough instruments to stop this practice. National parks officially accept local berry and mushroom pickers. In some regions, such as the Masurian Lakeland in Poland, this activity has changed from recreation into the main economic base for whole villages. Nevertheless, in Latvia and the Czech Republic this problem does not seem so serious, while in Hungary, with enormous geological attractions, mineral collecting appear more important.

Camping and picnicking are moderately popular activities in the region. Most problems occur from rubbish and fires due to picnicking, especially in protected areas and coastal zones, for example in Latvia. Camping is permitted only in indicated places, but it these restrictions are not very strictly applied. The severity of these activities has been evaluated as moderate, but it is related directly to infrastructure, discussed later. The problem of the low acceptance of regulations mentioned above also causes problems of illegal fire places in almost all countries of the region except Latvia and the Czech Republic.

The popularity of winter sports varies in Eastern European countries. Generally there is a low severity of conflicts relate to alpine skiing, ski touring, cross-country skiing, snowshoeing and snowmobiling in this region, except in the Czech Republic and Slovakia and southern Poland dominated by mountains – the High Tatras, Low Tatras, Great Fatra and other mountain chains. These conflicts also have a correlation with climatic features, and as the last decade had been less snowy than before, so the degree of conflict may change if users reduce in number.

The next group of recreational activities seem to be not so important in the Eastern European countries from the point of view of nature conservation. These are Nordic/power walking, geo-caching, wildlife/bird watching. Only in Slovakia are there some problems with poaching (collecting of bird's eggs, geo-caching) and nature conservation, while in Hungary there is a problem of people collecting antlers.

Of generally low severity and frequency of conflict are hiking, orienteering, horse riding and horse carriage driving, canoeing and swimming and climbing. Little differences occur between Eastern European countries in these particular activities. Hiking is not so frequent in most of the countries, only in Hungary and Slovakia is it really popular and a favourite pastime, especially in national parks, and there conflicts seem serious. Orienteering, canoeing and swimming are most popular in Slovakia, while traditionally horse riding and horse carriage driving is popular in Poland and Hungary. Except hiking all these activities cause few conflicts with the nature environment or nature conservation, but locally that may differ, e.g. horse riding off roads and trails can brings important consequences much more so than hiking or orienteering. Locations close to bigger cities where horse sports are more popular influences the scale of conflicts with the natural environment and with protected areas and nature conservation. Hills or mountains seem more susceptible to damage by horse riding than other places.

Climbing is very important in Slovakia and Hungary and only in rocky areas in Poland and the Czech Republic. The problems caused by climbing are focused of flora and fauna damage, but also with the disintegrating rock, particularly in limestone areas with unique land forms. Other than high levels of nature protection that almost eliminates or at least minimises this activity there is no proper solution in other areas to solve this conflict.

In the last decade cycling and mountain biking have arisen as sports of great popularity in most of countries. Only in Lithuania and Latvia is it not the case. Frequency of conflicts with nature protection are high in Poland and medium in the Czech Republic, Hungary and Slovakia. The problem is more complicated with mountain biking, especially in lowland areas in Central Poland, where natural conditions as well as a lack of special biking parks with proper infrastructure for freestyle biking leads to many changes to the landscape, and many urban forests have been damaged to fulfil bikers' needs illegally. Another problem is connected with protected areas where cyclists and mountain bikers need to use more trails that are available, so they use walking trails or even animal paths instead. As the popularity of theses activities increases, therefore solutions needs to include forest survey and/or rangers.

Water sports such as fishing and motorized water sports have different importance in particular country in Eastern Europe. Fishing is a very important sport in Latvia, quite important in Lithuania and Slovakia with frequent conflicts mostly with illegal fishing, while in other countries it is not significant at all. Enjoying motorized water sports is the most problematic activity for Lithuania, Poland, Slovakia and the Czech Republic. In Latvia water motorcycles have no special areas, and they are extremely easy to take even to remote places. In Poland it is quite clear which lakes are permitted for this kind of activity, but still in summer many other lakes are overcrowded by motorized water sports. Some instruments of nature conservation are helpful, such as nature reserves, but the category of regional park is not enough to protect lakeshore habitats.

6.5.3 Impacts and Disturbance of Other Societal Activities on Recreation

The Eastern European countries generally report rather low frequencies and severity of impacts and their conflicts concerning forest based recreation and nature tourism. Only a few elements have been rated as of high or medium importance. The following figure shows the frequency of impacts on forest recreation Fig. 6.8).

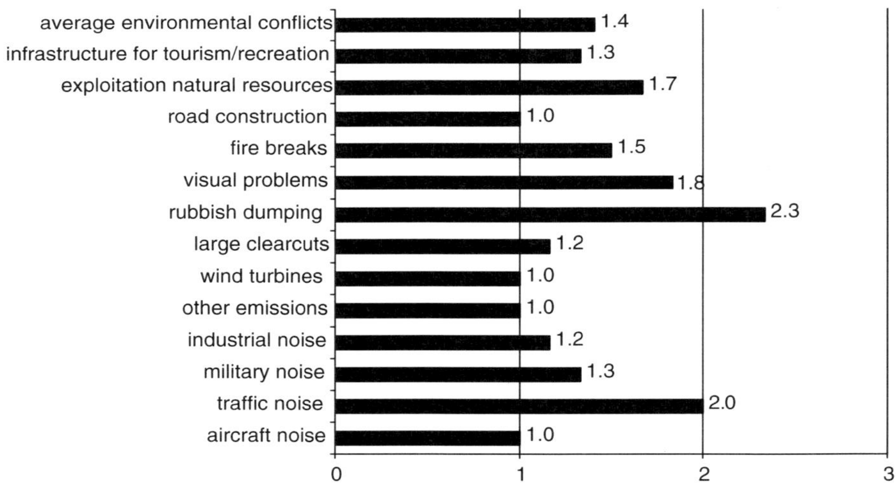

Fig. 6.8 Frequency of environmental conflicts in the Eastern countries (0=non-existent, 1=rarely, 2=medium, 3=frequently)

The most significant problem in all countries is rubbish dumping. The problem occurs not only in more popular areas such as urban forests, but generally across the whole country. The problem has its roots in three aspects. Firstly, in Communist times all forests belonged to the state, meaning to everybody, and the perception of looking after public property was quite low, next is the low level of environmental consciousness of society, and the third is the inefficient waste management system. The result indicates litter as the main problem across the whole region.

Medium severity has been rates for visual problems associated with damaged forests in all countries except Latvia, which was rated low. The forest most damaged by air pollution is located in the so-called "Black Triangle" – on the border of three countries – the Czech Republic, Poland and Germany. Some damage is connected with hurricanes such as the big wind calamity in the High Tatras in Slovakia in November 2004 or in the Masurian Lakes in Poland in July 2002 that destroyed large areas of forest. *Limantria dispar* and other insects caused some damage in Hungarian forests, the southern part of Slovakia and eastern of Poland and Lithuania. The frequency of conflicts caused by infrastructure for tourism (ski-lifts, cable cars) is also rated as medium in all Eastern European countries, except Hungary where it was rated as low.

Negative impacts of aircraft noise and wind turbines are quite low in the East European region, also there are no noticeable conflicts in this field. Wind turbines are quite infrequent in the region, except along the Baltic shore, but they are usually located far from forests. A similar range of problems concern military noise and other emissions (e.g. odour, dust), but there are some local differences. Generally they are rated as low, but in Slovakia other emissions have a medium scale of impact, while military noise is a problem in Hungary. In most areas military testing grounds are located in the forests, but rather far from accessible places. Nevertheless, each year in Poland there are accidents during the mushrooms season, and as a result few people may be wounded. This is again an effect of ignoring regulations and the temptation to collect rare mushroom species is stronger than common sense. Low levels of negative impact caused by industrial noise and road construction predominate in the Eastern European countries, only Hungary rating them as medium. Conflicts concerning road construction arise in Slovakian protected areas. This situation may change according to the priority of road construction for new members of the European Union.

Traffic noise, large clear cuts and fire breaks have different degrees of importance in Eastern European countries. Traffic noise varies from being a significant problem in the Czech Republic and Lithuania, less so in Poland and Hungary and infrequent in Latvia and Slovakia.

Large clear cuts are quite frequent in Hungary, of moderate occurrence in Latvia and Slovakia, and of low severity and frequency in the other Eastern European countries. In some countries the forestry policy has been changed in recent years and large clear cuts are not possible in Poland, the Czech Republic or Lithuania. Fire breaks do not produce conflicts in most countries; only in Slovakia are they rated as medium and high in Poland, where access to some places is limited. Rather low conflict levels caused by exploitation of natural resources that disturb the land cover (e.g. quarries) are reported in almost all the region. High severity and frequency is indicating in Hungary and medium in Slovakia.

6.6 Planning, Management and Monitoring

6.6.1 Planning for Recreation and Nature Tourism

Eastern European countries have an independent planning system for recreation and nature tourism in forests (Table 6.2). In Lithuania and Poland plans are prepared at regular intervals, while in other countries such as the Czech Republic, Hungary, Latvia and Slovakia, plans are made only in special cases.

Table 6.2 Availability of independent planning (legally required or optional) in the Eastern countries

	Cz	Hun	Lv	Lit	Pl	Sk
Independent planning for recreation and nature tourism in forests	Yes, special cases	Yes, special cases	Yes, special cases	Yes, regular	Yes, regular	Yes, special cases
Legal or optional requirement for forest owners	Required	Optional	Optional	Required	Required	Required

The planning process is similar only in Lithuania and Poland, where it is integrated in the forest inventory and forest management plan as well as into the regulations of national and regional parks. Other countries describe these procedures as follows:

- Czech Republic: the Ministry of Regional Development, regional governments and municipal offices and the administration of nature conservation are responsible for planning.
- Hungary: planning is usually undertaken by forestry personnel, sometimes by landscape architects. There are no clear regulations regarding forests recreation planning. Near to cities and especially near to Budapest recreation planning (as well as the management of recreation infrastructure) is more intensive.
- Latvia: plans are prepared on a voluntary basis in state forests and by Riga City Forest Agency, as well as in protected areas by their administrations – mostly in areas of high landscape quality or places with a high concentration of visitors.
- Slovakia: regional offices (self-governments) and forestry enterprises prepare the programmes for recreation and tourism development. The government committee for tourism co-operates with them and co-ordinates tourism and recreation development across the whole of Slovakia.

The planning of recreation and nature tourism in forests is a legal requirement for forest owners in Poland, Slovakia, the Czech Republic and Lithuania. In Hungary and Latvia it is only optional. There is literature helpful to direct planning and to manage recreation and nature tourism in forests in the form of handbooks, guidelines or other instruction frameworks available for all countries except for Latvia. However, the literature of each country is very wide-ranging. There are only general recommendations and planning principles presented in the handbooks of landscape planning and forestry in Hungary, Poland, Slovakia and the Czech Republic. Literature is more specific in Lithuania – handbooks and yearbooks of forestry exist as well as a framework for the recreational objects in Forest Enterprises – surveys of the recreational objects and the implementation of special projects, for example recreation for disabled people in the forest.

6.6.2 Planning Steps

Across all of the Eastern European countries with the exception of Latvia there are only a few similar steps undertaken for planning forest recreation: inventory of ecological conditions, definitions of visions and goals and participation in the definition of goals (Table 6.3).

Table 6.3 Steps that are undertaken for planning forest recreation at regular intervals in the Eastern countries

Steps in planning forest recreation	Countries
Inventory of ecological conditions	CZ, HUN, LIT, PL, SK
Inventory of social conditions	CZ, LIT, SK
Analysis of conflicts	CZ, SK
Definition of vision and goals	CZ, HUN, LIT, PL, SK
Participation in the definition of goals	CZ, HUN, LIT, PL, SK
Definition of standards	CZ, LIT, SK
Deduction of measures	SK
Participation in the deduction of measures	LIT, SK
Monitoring	CZ, LIT, PL, SK

In addition, an inventory of social conditions and an analysis of conflicts are prepared for all except Poland. Almost all steps are practiced in Slovakia and Lithuania. Standards are also defined in the Czech Republic, while the deduction of measures is a step taken in Hungary. Monitoring is not a part of the planning process in Hungary. In Latvia there are no guidelines for such planning available and the steps are defined case by case.

The availability of data for planning is generally quite satisfactory in all countries; however, the different sources are often – also within a country – not comparable or their availability from different institutions is complicated. Data can be obtained from national surveys in the Czech Republic, Hungary, Lithuania and Poland. There are data from regional surveys in Hungary, Latvia, Poland and Slovakia. Data is collected by the forest administration in all countries, except for Lithuania. Other sources, especially for planning purposes are available in Hungary, Latvia, Poland, Slovakia, the Czech Republic and Lithuania.

Public participation in the definition of visions and goals seems to be quite common in Eastern European countries, except for Latvia. As such planning is not common in Latvia, public participation is not relevant. Most of the stakeholders are involved in the planning processes in all Eastern European countries. Mushroom and berry collectors do not take part in the definition of visions and goals – this group is usually not affected by the plans as access to the forest is free in the Eastern countries and they use particular parts of forests only occasionally. In Poland hiking clubs and sport associations are excluded, but they are responsible for planning and maintaining tourist trails. In Lithuania the lack of local residents as well as hunting and fishing representatives is significant. Hungary, Lithuania, Poland and Slovakia pointed to NGOs as important participants in planning.

Monitoring of plans is carried out at different time intervals in the each country. In Hungary it is prepared at quite long intervals, approximately in 10–15 years, while in Slovakia it is done every 10 years or when indicated as being necessary. In the Czech Republic monitoring takes place every 1–5 years. In Poland monitoring is carried out less than once in five years and in Lithuania at irregular intervals, only when it is seems as necessary.

6.6.3 Integration of Recreation and Tourism in Forest Planning and Monitoring

Each country was asked to evaluate the integration of forest recreation and nature tourism into the planning and monitoring of that country's forests. The overall evaluation was rated as being good by the Czech Republic, Hungary, Lithuania, Poland and Slovakia. Latvia assesses its planning as satisfactory, which corresponds with the situation described for the country.

Generally it seems that the level of priority given to forest recreation and the quality of forest recreation planning could be better in every country. This however does not mean that a more intensive infrastructural development would be necessary in the the forests of these countries since the levels of use are not so intensive as in the Atlantic region, for example, and the types of activities and expectations of the users are also not so high.

All the countries demonstrate regional differences regarding the treatment of forest recreation and nature tourism. Hungary has big differences partly because of the uneven distribution of forests – due primarily to differences of the climate and relief – and partly because recreation demand is greater near the bigger cities. Poland, Slovakia, Latvia and Lithuania also have considerable differences due to two diametrically opposed reasons – the first being the easy access to forests in and around urban areas and the second being the restrictions placed on access to protected areas or other remote areas. In Lithuania these differences are becoming less extreme due to the development of countryside tourism. In addition the different economic conditions of owners as well as lack of knowledge and experience are factors in Latvia.

6.7 Recreational Infrastructure

The provision of basic and additional recreational infrastructure was evaluated across the region. Figure 6.9 shows the ratings of this for each country. Poland is rated as the most unsatisfactory, urban areas being the worst equipped. By contrast, in Hungary, the Czech Republic and Slovakia, the basic infrastructure generally seems to be in a better condition. Lithuania and Latvia stand out the most, as leaders of the provision of basic recreational infrastructure in the Eastern Region. This result corresponds only partly to the evaluation of particular infrastructure elements, where Lithuania had been assessed much better than other countries in the region. By contrast, the high rating of Latvia does not find correspondence with provision of specific elements, and generally Latvia is in the lower ranks of infrastructure provision despite what the rating states.

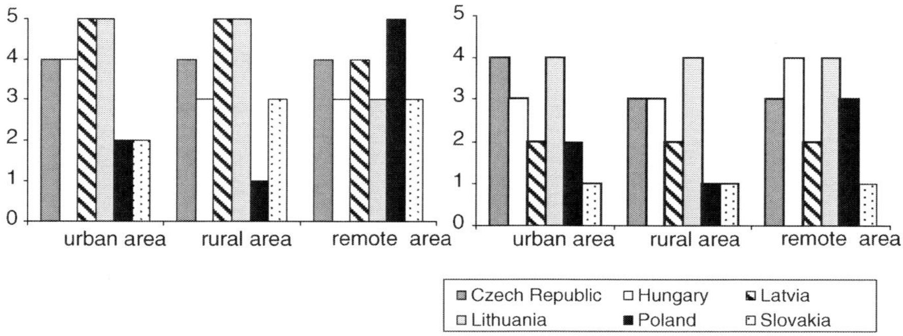

Fig. 6.9 Supply of basic infrastructure (*left figure*, **a**) and additional infrastructure (*right figure*, **b**) in the Eastern countries (no bar = non-existent, 1=unsatisfactory, 5=very good)

In the Eastern European region the amount of additional infrastructure is rather low. In all countries the rural areas are rated as the most unsatisfactory, and remote areas seems to be in a better condition. Lithuania stands out as being almost excellent in rural areas compared to the rest of the countries.

Generally, in Eastern European countries there are no particular standards for the provision of access. Benchmarks are only used in Slovakia for recreational management in all locations – urban, rural and remote – and also in remote areas in Lithuania.

6.7.1 Additional Facilities

The results of the additional infrastructure comparison indicate that some elements are more frequently found in urban areas, such as huts, fire places, benches, viewing platform, play grounds and rubbish bins. The occurrence of additional infrastructure in urban, rural and remote areas in the European countries is presented in Fig. 6.10.

These elements are more frequently used close to cities in most of the countries of the region. Latvia and Czech Republic only rarely provide such elements of infrastructure generally, with urban areas not being treated much differently. The

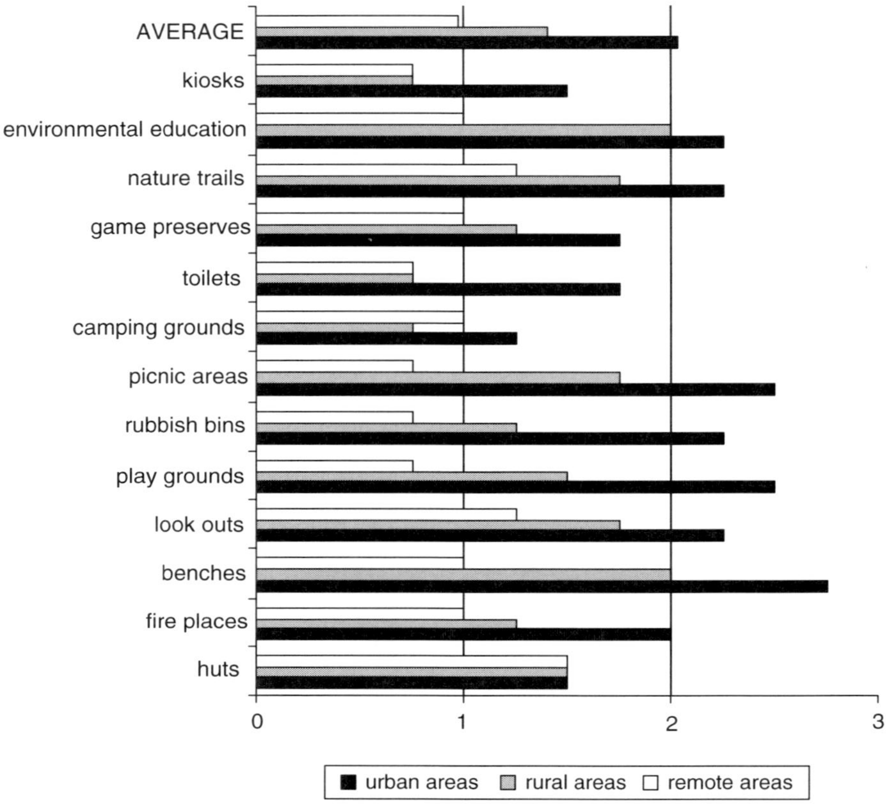

Fig. 6.10 Occurrence of additional infrastructure in urban, rural and remote areas in the East European countries (0=non-existent, 1= rarely, 2=reasonable, 3=frequently)

overview of infrastructure provision shows a lack of playgrounds in rural and remote areas. Viewing platforms and rubbish bins are also quite rare in remote areas. This shows the paradox of maintenance in remote areas. High natural values seem to promote activities such as wildlife watching and nature observation but the infrastructure does not appear in the proper place. A lack of rubbish bins in remote areas may creates threats if people leave a lot of rubbish lying around. However, there can be a problem with rubbish bins natural areas if wildlife can get into them and eat the rubbish or spread it around. Bins need to be designed so that people can use them but so that wildlife cannot get into them.

Another group of infrastructure elements are less frequently found in the region compared with other regions regardless of the type of area (urban, rural or remote). These are picnic areas, camp grounds, toilets, game preserves and kiosks. Among these elements picnic areas and camp grounds are relatively more common than the others. This shows that there are some infrastructure elements needed more than others. The need can also arise from traditional uses of the forest. For example in Poland it is officially forbidden to camp in the forest, but in fact camping is not very restricted; also, since the management policy of remote areas is to avoid camping on sensitive nature sites, most campgrounds are therefore located outside the forest. Lithuania emerges as the most equipped country in the Eastern European region, followed by Slovakia and Hungary.

Finally, there are two infrastructure elements that demonstrate some changes in management practices. Nature trails and sport trails as well as environmental education facilities are much more frequent in all the countries and in all area types. That shows that this generally rather new educational orientated management has developed quickly in the Eastern European region. Slovakia and the Czech Republic have the highest quality network of marked tourist trails through forests in Europe. The significant lack of toilets presents a potentially significant problem in all the countries of the region. Pollution in busy areas can be high where there is a lack of toilet provision.

According to rate of the infrastructure no countries consider the forests to be over-equipped for recreational needs. Lithuania seems to be better equipped than the other Eastern European countries, and the only one to point to too much infrastructure but then only in urban areas.

6.7.2 Quality of Infrastructure

The quality of infrastructure was rated as good in most countries (Fig. 6.11). Only Poland rated this as unsatisfactory in all aspects. Excellent quality of infrastructure is only rated so in rural areas of Hungary, which is because it is a newly developed area where the infrastructure is quite recent.

The main problem of busy areas such as near urban areas is that infrastructure is often damaged quite soon after installation. Latvia points to a lack of experience in design and in the safety requirements of new infrastructure, which is made from natural materials such as wood. Latvia also indicates that in national parks and other protected areas with good administration the infrastructure is excellent. In Lithuania

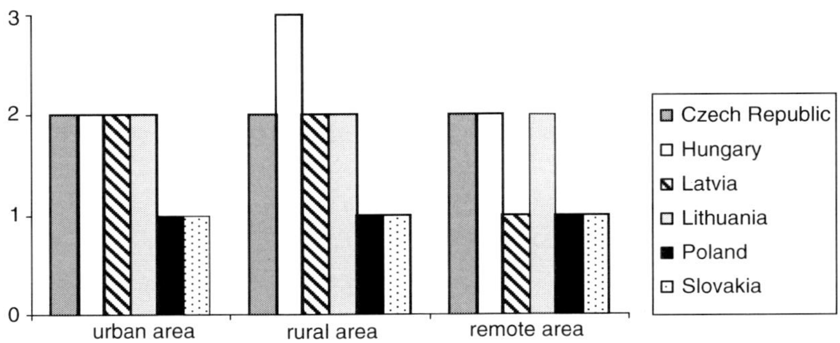

Fig. 6.11 Quality of infrastructure for recreation and nature-based tourism (no bar = non-existent, 1=unsatisfactory, 3=excellent)

the quality depends on the land owner or manager. In Poland and Slovakia foresters are responsible for the design and safety of infrastructure, but hiking and cycling clubs are responsible for marking, building and maintaining the trails. There is no requirement for landscape architect involvement in the design process.

6.8 Approaches to Visitor Management

Different management actions are undertaken to improve quality of forest recreation and to reduce conflicts. They are usually divided into three main groups:

- *restrictive management actions* – by imposing different limitations on specific activities to improve the forest recreation quality and to help to eliminate conflicts – the effect of these actions are immediate;
- *soft management actions* – such signposting to guide visitors around the forest, to show the location of the most attractive areas but also to keep people away from sensitive places and to provide attractive infrastructure. This approach to recreation management is more sophisticated and expensive but the effect of these actions are long term and public acceptance is higher than for restrictive actions;
- *facilitating management actions* – indirect methods to improve the quality of the forest recreation experience such as visitor guiding and environmental education which helps to encourage visitors to behave more sensitively; it also promotes more appropriate activities and eliminate threats and conflicts. The effects of these actions are also long term.

The following figures present the application of management actions in the Eastern region (Fig. 6.12) and the level of acceptance for the various management

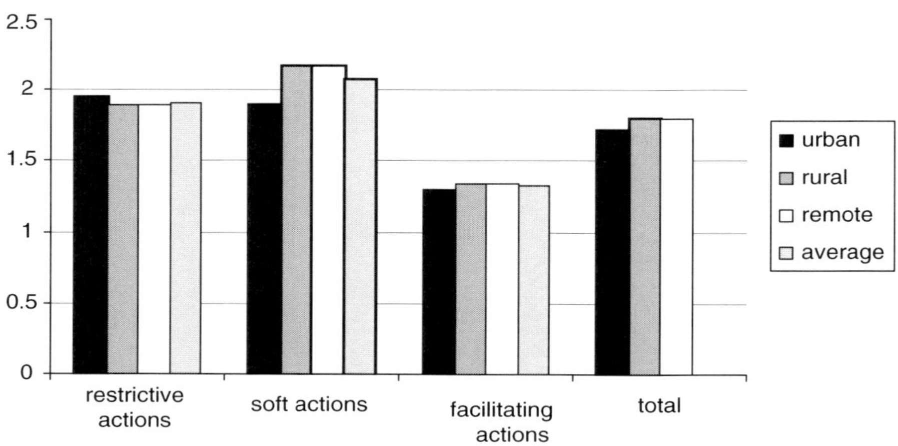

Fig. 6.12 Occurrence of management actions in the Eastern region, in general and in urban, rural and remote areas (0=non-existent, 1= rarely, 2=regularly, 3=frequently)

Table 6.4 The level of acceptance in the Eastern region for the various management actions (high >= 2.5, medium > 1.5 and < 2.5, low =<1.5)

Acceptance by the public	Restrictive management actions	Soft management actions	Facilitating management actions
High		Attractive infrastructure Educational infrastructure Signposting responsible visitor behaviour	
Medium	Entrance fee Limiting visitors' entrance		Guiding thematic tours Guiding ranger systems Training special user groups Training commercial providers
Low	Rules and regulations Prohibitions Marking protected areas/tabuzones Barriers and fences		

actions (Table 6.4). Soft actions seem to be the most frequently applied but there are regional differences and also variation by types of recreational areas, being less used in urban areas but more so in other areas. Restrictive actions are applied more equally at all types of recreational areas.

The general picture seems to show that of the management actions, the soft types have the strongest degree of acceptance and approval by visitors. Restrictive management actions, usually the easiest to implement, are only accepted at a low or medium level in the Eastern Region.

6.8.1 Restrictive Management Actions

One of the most popular management actions among Easter European countries are restrictive ones, particularly the use of rules and regulations. Only in Latvia are prohibitions assessed as being rarely applied, but in all other countries these frequently dominate. Unsurprisingly their acceptance by the public is relatively low, only in the Czech Republic, and Hungary are they rated as medium. Less frequently applied but also a significant method of recreation management are the use of barriers and fences. The popularity of these approaches derives from communist times, when it was much easier to forbid activities, than to permit them. Other actions are generally not popular in Eastern European countries, but there are some regional differences. Entrance fees are common in the Czech Republic, while marking protection areas and no-go zones is common in Poland and Slovakia. Limiting visitors' access is the least used method among all the countries of the region, but its' public acceptance is high in Poland and Slovakia and medium in Czech Republic. This last technique needs a high degree of organization with clear rules and strong financial support – compared with the use of entrance fees, this is not a self-funding method. Generally, the degree of public acceptance of this group of management actions is rather low. The Czech Republic and Hungary, and to some extent Latvia, accept these restricted methods in some aspects at a medium level but in other countries such as Lithuania, Slovakia and Poland the degree acceptance is mainly low. In many cases the problem concerns the communication between visitors and forest managers. A typical example is an information board presenting a long list of regulations, often difficult to read and written in legal language list so also difficult to understand and therefore ignored. This shows that it is easier to post the rules than to explain them; the degree of public acceptance shows their effectiveness.

6.8.2 Soft Management Actions

Soft methods are applied in different ways in Eastern European countries. It seems that each country applies a different method to improve visitor management. Hungary gives the highest value to all three types of action, while other countries have focused on some of them. Improving areas with new attractive infrastructure is not only a Hungarian but also a Lithuanian approach. Similarly rated highly is the use of new infrastructure for ecological education, but in this case Poland, the Czech Republic and Slovakia also use actions. Latvia rates this as being used less. Signposting for visitors is not used in the Czech Republic and Latvia but is used in the other countries. The Czech Republic and Latvia tend not to use this soft group of actions. Generally this group of methods has the highest degree of public acceptance, particular the relatively less used approach of improving areas with new attractive infrastructure.

6.8.3 Facilitating Management Actions

Facilitating actions tend to have the longest-lasting influence on visitor management. They need significant financial support and takes time for them to take effect, so this group of techniques is the least likely to be applied in the Eastern European group of countries. Guiding visitors and using thematic tour groups is mostly used in Lithuania and, to a lesser extent, in Hungary. Poland and Slovakia are experienced with this method, but mostly use it in remoter areas. In Latvia these actions are applied more often in rural areas. Providing special training for user groups is a Slovakian specialization and training for commercial providers are infrequently used but considered for remote areas. Facilitating actions have a medium or high level of acceptance.

6.9 Future Trends and Challenges

All the Eastern European countries are in a forest ownership transformation process. Private ownership is increasing, which may in some cases influence future forest accessibility and in consequence initiate conflicts in the field of recreation. Also three countries with the highest forest cover in the region: the Czech Republic, Slovakia and Latvia, still promote timber production as the main function of forests. The Polish approach with a multifunctional model of forest use is an advanced, modern idea that can present some challenges for other countries of the region.

Among the different types of forests – urban, rural and remote – the most important recreational role is fulfilled by forests around cities, usually used for short duration visits. They are the most equipped type of forest compared with the rural or remote forests in the region, but also under greater stress from visitors. Another critical group in Eastern European countries is remote forests that have been enlarged recently, especially in terms of increased protection (e.g. Natura 2000 areas); here a lack of infrastructure is the first problem that can lead to degradation, mostly of the biological value of a site. In the region forests in rural areas fulfil the traditional role as a source of berries and mushrooms.

The next trends concerns conflicts. The most important conflict in the region is between visitors and nature protection. In Eastern Europe it may bring a loss of certain biological values in some forests. Changes to this situation require a transformation of the ecological consciousness of society and a deeper understanding and acceptance of the need for protection. Another important challenge is modification of the perception of the forest as a more valuable element of the landscape. This goal seems to be long term, but it may bring different benefits such as a reduction of rubbish dumping – the most important problem in all the Eastern countries.

Another trend with a long term perspective is the introduction of nature trails and educational orientated management that increases ecological awareness among society. However, the most important factor affecting forest recreation in Eastern Europe remains the lack of basic infrastructure. This step also affects the major conflict between visitors and nature protection mentioned above.

A final important trend in contemporary forest recreation in this part of Europe is the changing model of different sport activities and in particular the increasing role of cycling, mountain biking and motorized sports such as off-road vehicles and quad bikes. Also, new types of recreation activities with a sport connection developing in the forest are "Go Ape" type aerial ropeways, roller coasters, summer snowboarding, mountain grassboarding, three-wheeler downhill going and others. They can cause a new set of threats to the forest in spite of strong rules and regulations. These regulations do not work without public acceptance. The result of this report indicates that soft/incentive-based methods are more effective in recreation management, even if the results will not come immediately.

Chapter 7
Cross-European Comparison

Birgit Elands and Veronika Wirth

7.1 Introduction

The aim of this chapter is to draw a cross-European comparison of the differences and similarities of forest recreation and nature tourism. The main results of Chapters 2–6 will be analysed and discussed in a broad European context. The analysis will be directed at comparing and interpreting similarities and differences between regions and countries. Table 7.1 gives the overview of involved countries per region.

Table 7.1 List of countries in the five European regions (abbreviations in brackets)

European region	Involved countries
Atlantic region	Belgium (bel), Denmark (dk), Iceland (ic), Ireland (ir), the Netherlands (nl) , United Kingdom (uk),
Nordic region	Finland (fin), Norway (nor), Sweden (swe)
Central European region	Austria (at), France (fr), Germany (ger), Switzerland (ch)
Eastern region	Czech Republic (cz), Hungary (hun), Latvia (lat), Lithuania (lit), Poland (pl), Slovakia (sk)
Mediterranean region	Bosnie-Herzegovina (bos), Croatia (cro), Cyprus (cy), Greece (gre), Italy (it), Portugal (por), Serbia-Montenegro (srb)

The chapter is structured as follows: First, forests in different types of societies in Europe are compared; forest cover, population density and forest ownership patterns are analysed at a European level. Next, legislative frameworks and functions of the forest as well as forest managers' knowledge of recreation are compared across Europe. Conflicts related to recreation are portrayed in the different regions; planning solutions, recreational infrastructure and visitor management are also compared. The chapter closes with conclusions and an outlook on remaining problems and future challenges for forest recreation.

B. Elands (✉)
Wageningen University & Research Centre, Forest & Nature Conservation Policy Group, Droevendaalsesteeg 3, 6708 PB Wageningen, The Netherlands
e-mail: birgit.elands@wur.nl

U. Pröbstl et al. (eds.), *Management of Recreation and Nature Based Tourism in European Forests*, DOI 10.1007/978-3-642-03145-8_7,

7.2 Forest Cover, Development and Ownership

7.2.1 Forest Cover

There are significant differences in forest cover throughout Europe. These differences are to a large extent dependent on historical development, on population densities, land use pressures and on climatic conditions. Figure 7.1 shows the forest cover of each European country and contrasts the different regions. Finland has the highest percentage of forest cover in Europe (75% of the total land area), whereas Iceland has the lowest forest cover (only 1.3% of the total land area). The average proportion of forest cover in the countries covered by the investigation in Europe is 31.7%.

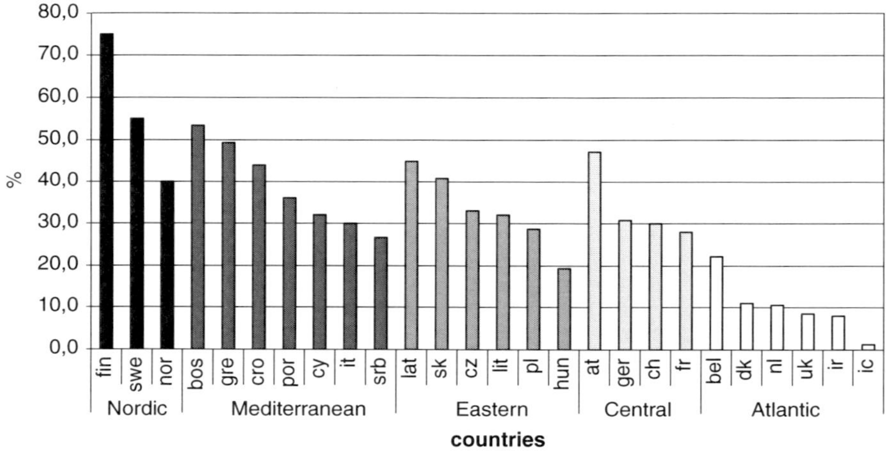

Fig. 7.1 Forest cover (%) in 26 European countries

The highest percentage of forest cover can be found in the very Northern and Southern parts of Europe (see Table 7.2). The Nordic region on average has the highest percentage of forest cover (57%); the Mediterranean region has an average forest cover of 39%, whereas the Atlantic region has the lowest forest cover (10%). The average forest cover in the Central region and Eastern region is around the average forest proportion for Europe as a whole (Central region: 34%, Eastern region 33%). In both of these regions there is a tendency for forest cover to increase. The high proportion of forest cover in the Mediterranean region is partly a misleading statistic because the definition also includes pre/post-forest shrubby vegetation (maquis-like vegetation) and similar forest types which do not occur in other countries and which would not be recognizable to most people as "forest". The relative importance of maquis-like vegetation on the total vegetation cover ranges greatly

Table 7.2 Forest and population characteristics of European regions

European region	Forest Cover	Forest ownership * Public/state	Private	Population density	Amount of forest (ha) per inhabitant
Atlantic region	10%	52%	48%	210	0.13
Nordic region	57%	22%	75%	18	3.26
Central region	34%	44%	56%	156	0.26
Eastern region	33%	58%	39%	94	0.47
Mediterranean region	39%	59%	41%	113	0.39

∗ Public/state and private forest ownership summarized are not always 100%
Statistics: forest cover: $eta^2 = 0.702$ ($p < 0.001$), forest ownership (nordic/central vs. other region): $eta^2=0.215$ ($p < 0.05$) and $eta^2=0.215$ ($p < 0.05$), population density: $eta^2=0.311$ ($p < 0.10$), amount of forest/inhabitants: $eta^2=0.937$ ($p < 0.001$)

among the countries. For example Cyprus shows 32% forest cover of which 18.5% is high forest and 13.5% is shrubby vegetation (maquis forest).

An important factor for forest recreation and nature based tourism is the relationship of forest cover to population density. Again, there are large differences throughout Europe. Figure 7.1 and Table 7.2 show that the Atlantic region is the most densely populated, closely followed by the Central region. The Nordic region has a much lower population. If the amount of forest and other wooded land per inhabitant is calculated, the differences between the regions become even larger. Nordic inhabitants have on average about ten times more forest available to them than the inhabitants of the other European regions. The most urbanised are the Atlantic and Central regions. In this context not only the amount but also the distribution of forests in a country has to be considered: how easily the population can reach forests for recreation. However, the results correspond well with each other.

7.2.2 Ownership

Forest ownership is another important matter affecting forest recreation. Forest ownership patterns differ widely throughout Europe. In the Nordic region private forests clearly dominate whereas in the other regions the proportions of private and public/state forest are more equal as shown in Table 7.2. However, the forest ownership patterns in the respective regions are not at all homogenous (see Chapters 2–6). The patterns are influenced by factors such as history and legal traditions that are very specific to each country. One of the most significant historical influences is the hunting tradition. Former royal hunting forests often form the backbone for recreation and nature based tourism in many Atlantic and Central European countries.

7.3 Legal Background, Public Access and Forest Functions

7.3.1 Historical Background of Public Access

National legislation provides the basis for public access to and recreational uses of forests. There is a large variety of legal regulations throughout Europe. They are strongly influenced by historical traditions and forest ownership. Roman law traditions stress the rights of the land owner. The land owner has the full power of use and disposal on his land. In countries with a Roman law tradition ownership still influences the accessibility of the forest. Private land owners may restrict public access to their land. In the Germanic law traditions there is a staged property right; this means several people may have property rights on one piece of land. Linked with this ownership were a number of duties. Anchored in the Germanic law traditions is also the idea of common land that may be used by everybody. Countries where the Germanic law traditions have a long tradition have a comparatively free access right (cp. Muhar 2004, Pinterits 2003). A question which also influences access for recreational purposes is the question of liability of the land owner. For example in Austria according to the General Terms and Conditions Act the land owner is liable if a person gets injured because of wantonly negligent road conditions.

The access rights in Europe are historical individual rights, which apply not only to locals, those who live in the country, but also to foreigners. The Nordic countries as well as Iceland are characterised by the greatest freedom of access to forests in European terms. There is a right of free access to all natural areas (including forests), or more precisely "uncultivated land". In Norway and Sweden it is called "allemansretten" which can be translated as "everyman's right". This right also permits access to private land. This is the most liberal regulation in the European Union. It does not mean, however, that all recreational behaviour is allowed. Access is free only if visitors behave in a responsible way with respect to the landowner's property, the environment and other visitors. Apart from this, free access does not include the use of motorised vehicles. Finally, this free access does not necessarily apply to newly planted forest areas.

Several Central European countries, such as Austria, Germany and Switzerland, also have a system of mainly free access to forests for recreational use. There is no influence by the owner as to whether land is accessible of not. There may be restrictions on the use of certain recreational equipment. Except for the Northern parts of Germany there is no requirement to stay on trails if the activity does not have an impact on land use and agricultural products.

The Atlantic countries have or had until recently the most restrictions on access to private forests. In the UK the land owner may influence public access to their land. However, there is a right of way on defined roads and paths. Under the Countryside and Rights of Way Act 2000 there is free access, the so called "right to roam" on certain uncultivated areas. In France, the Netherlands, Ireland and most of the Eastern and Southern European countries the regulation of public access depends on ownership. In general, this implies that public forests (including state forests) are freely accessible and the access in privately owned forests can be restricted.

Due to an increasing pressure by tourists and recreationists around cities and densely populated areas the legal conditions have been changed in different ways. A first exception relates to urban pressure. In those countries with a high urban pressure, and in some cases also a low forest cover, there are several initiatives towards either introducing a statutory right of access on foot or encouraging private owners, through financial inducements, to make their property accessible to the public. Both in Denmark and Flanders, public access to all forests - at least by foot and during daytime - is incorporated in national legislation, whereas the Netherlands has created a system in which private forest owners are subsidized for providing public access. New more liberal regulations have recently been established in Scotland where private forest land is now accessible to the public.

Other reasons for limiting public access at least temporarily are in order to protect flora and fauna, in the nesting or hunting season or for safety reasons (e.g. wood harvesting).

7.3.2 Effect of Legislation

The survey asked each expert to indicate if any legislation exists in their country that enables forest owners to affect public access for recreational uses (e.g. recreation activities that do not require any special equipment, such as hiking, cycling and collecting mushroom and berries). Almost all countries, except for Greece, Iceland, Ireland, Norway, and Sweden, confirm this to be the case. Table 7.3 shows that there are clear regional differences: all Central and Eastern countries have legislation to regulate public access, whereas only one of the three Nordic countries has. In the Central countries there is free access to the forest without regards to ownership, except in France, where restrictions may only be implemented due to nature conservation issues. There are restrictions for motorised vehicles and there may be restrictions for different type of sports.

Table 7.3 Existence of legislation affecting public access for recreational uses (%)

		Restrictions apply to (N=21)		
Region	Existence (N=26) Yes	Private forests	Public forests	State forests
Atlantic	67	100	50	50
Nordic	33	0	0	100
Central	100	25	100	100
Eastern	100	67	67	100
Mediterranean	86	83	33	83
Total	81	67	57	86

Not surprisingly, most legislation applies to state forests, although legislation also exists for many private and public forests. In the only Nordic country with legislation (Finland) it applies exclusively to state forests. Again, the Central and

Eastern countries have on average the most extended legislation in terms of a variety of ownership.

A second point concerns the existence of laws affecting the recreational use of forests. Two groups of countries confirm that there are either general directions (N=9) or both general and specific directions (N=10). A limited number of countries (N=5) have several laws pertaining to recreational uses. Only in Iceland are there are no directions or laws regarding recreational use of forests.

Figure 7.2 shows more specifically which types of recreational use are regulated by law. It can be concluded that in almost all countries laws exist that affect both public access and the collection of non-timber forest products (NTFP's). This is also largely true for the regulation of sport activities. Finally, several countries indicate that other recreational uses are covered in legislation as well (e.g. dogs, litter, fire precautions).

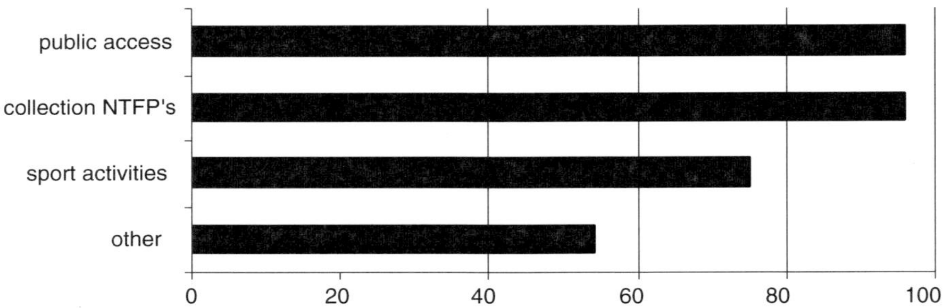

Fig. 7.2 The existence of laws affecting different recreational uses of forests (%)

7.3.3 Functional Classification Systems

At a national level certain classification systems can be used to indicate the importance of forests and nature areas for the different functions they fulfil for society, for example recreation, biodiversity, cultural history, protection, etc. These classification systems may be incorporated into forest laws or similar ordinances. Within this system forests can classified as recreational forests, urban forests, forest parks, etc., having recreation as the primary or sole management objective. Many countries (73%) are acquainted with such a functional classification for recreation; this is the case especially in the Nordic and Eastern regions (100%). The importance of such a classification in the spatial planning system, however, is not uniformly spread around the regions: for the Nordic and Eastern countries it is very important (67 and 83% respectively), whereas in only a few Atlantic and Central countries is it very important (17 and 25% respectively).

The nature conservation legislation of many countries contains several categories of protected areas. Some of those categories are likely to be found in forested

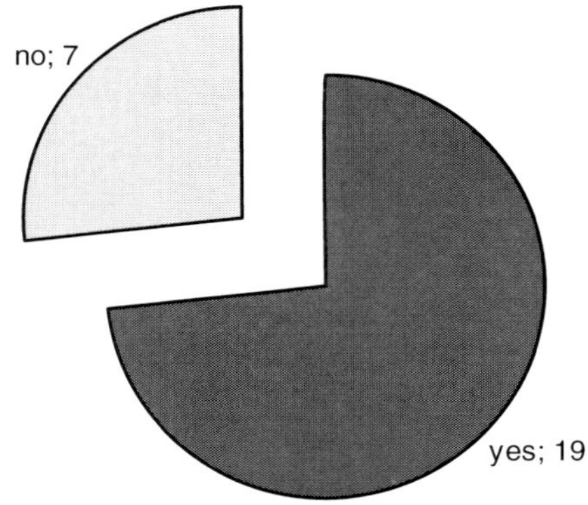

Fig. 7.3 The existence of a functional classification of forests as recreational forests or forest parks (countries)

areas and recreation may play an important role in it. The following categories of protected areas exist in most countries:

- *National parks*, natural areas with, nature conservation, education recreation and research goals (96%)
- *Protected areas for landscape conservation*, areas or regions which are being protected for their typical natural scenery (96%)
- *Nature parks*, a model landscape, usually cultural landscape, which is protected to enhance recreation and tourism (69%).

Apart from these three major categories, a majority of the countries also mentioned another type of protection. There are no differences between the regions concerning national parks and protected areas for landscape conservation. However, nature parks occur in the Central and Eastern countries and rarely in the Atlantic countries. There is one important difference between the regions: Central and Eastern European countries identify on average the most categories of protection (3.8), whereas the Atlantic countries identify the fewest (2.3). This difference might be caused by cultural differences as both regions have a tradition of strict regulations and classifications.

7.3.4 The Importance of Recreation as a Forest Function

The national experts were asked to describe the importance of the different forest functions and the relative importance of the recreational function in their countries (Fig. 7.4). In two countries all forest functions are perceived as equal (Denmark and Germany). Recreation is important in all forests in Finland and Norway. In four countries both recreation and biodiversity are important (Cyprus, Denmark, Netherlands, United Kingdom). Most countries stated that recreation is locally more important: near cities and/or in tourist regions or in dedicated areas

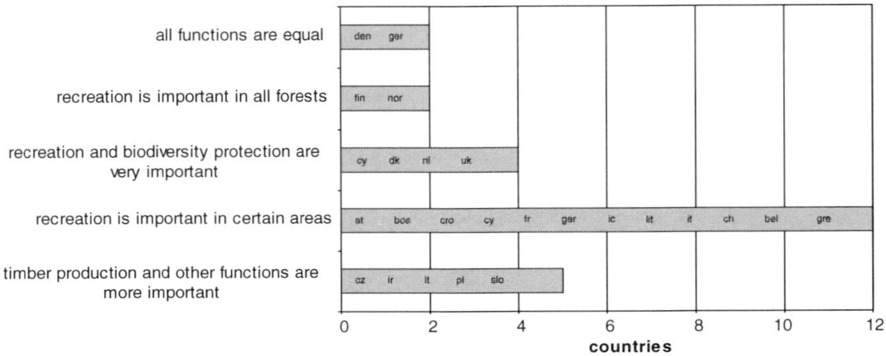

Fig. 7.4 Presents the significance of the recreational function in European forests. Statements are analysed by country (multiple replies are possible) (No answers from hun, lat, srb, swe, por)

(Austria, Belgium, Bosnia, Croatia, Cyprus France, Germany, Greece, Iceland, Italy, Lithuania and Switzerland). Five countries state that timber production and other functions are more important (Czech Republic, Ireland, Italy, Poland, Slovakia).

If we analyse the functions at a regional level it seems that in the Nordic regions recreation is important in all forests. In the Central regions recreation is locally important (near cities and in tourist regions). In the Eastern region other functions are usually more important.

7.4 Forest Managers' Knowledge of Recreation and Tasks Regarding Recreation in Forests

Forest managers' knowledge of planning and management of forest based recreation is an important requirement for a good consideration and management of forest recreation. In Europe, in 18 out of the 26 countries in the survey recreation is considered to be part of the basic knowledge and skills of forest managers.

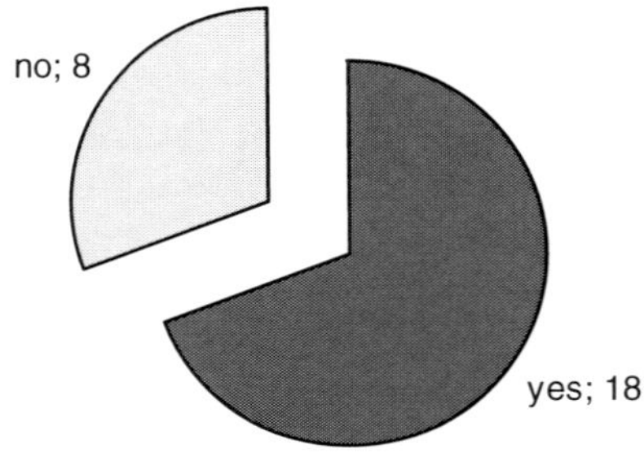

Fig. 7.5 Is knowledge about recreation considered to be part of the basic knowledge and skills of forest managers? (countries)

In Europe, there are significant differences between the regions regarding forest managers' knowledge of recreation (Table 7.4). In the Nordic and in the Central region knowledge about planning and management of forest based recreation is considered part of the basic knowledge and skills of forest managers whereas in the Atlantic and Eastern region this is only partly the case. In the Mediterranean region, forest managers are only in some countries supposed to deal with recreational aspects. Forest managers in Europe acquire their knowledge mostly during their formal education but working experience also plays an important role - in the Atlantic region working experience is the most important basis for forest managers' knowledge of recreation.

Table 7.4 Evaluation of Forest managers' education and working experience

	Basic knowledge about planning and management of forest recreation (%) *	
Region	Yes	No
Atlantic	67	33
Nordic	100	0
Central	100	0
Eastern	83	17
Mediterranean	29	71
Total	69	31

* Cramer's V = 0.592 (p=0.05)

Figure 7.6 shows the evaluation of forest mangers' knowledge about planning and management of forest based recreation in the different countries in Europe. Both education and working experience are evaluated.

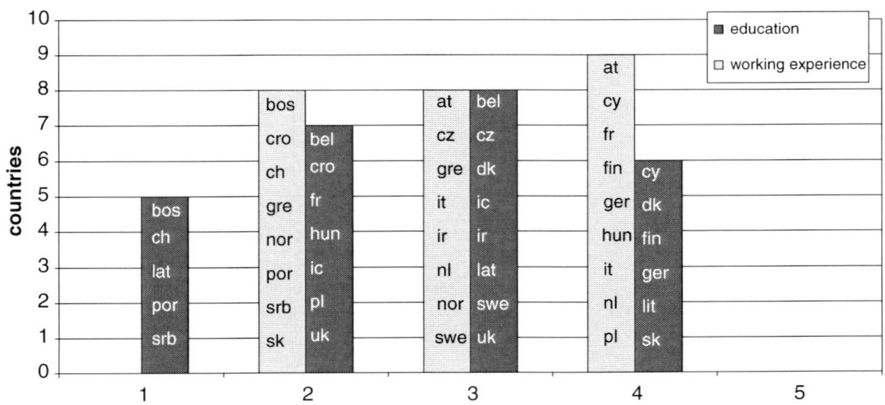

Fig. 7.6 Evaluation of knowledge about planning and management of forest based recreation obtained by education and working experience on country level (1=unsatisfactory, 5= excellent)

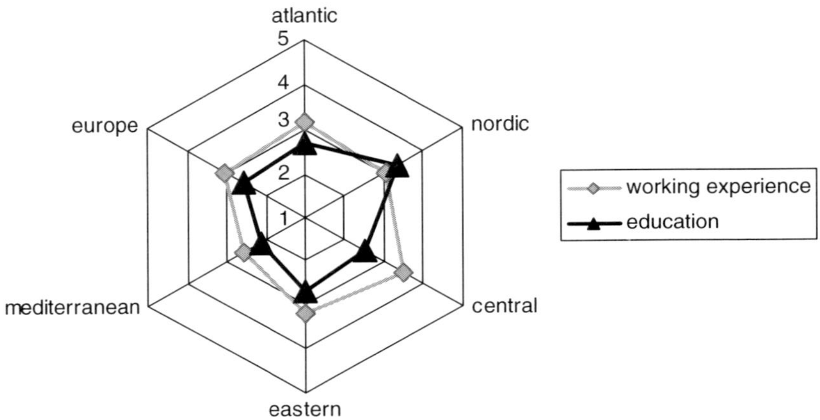

Fig. 7.7 Evaluation of knowledge about forest based recreation obtained by education and working experience at European and regional level (1= unsatisfactory, 5=excellent)

At the European level the experience gained through a working experience (mean=3.0) is considered to be more important than the quality of education (mean=2.6) (Fig. 7.7). The results show that in the Nordic countries the knowledge forest manager obtains from education is considered to be the highest. The Nordic countries have national surveys and special research on recreation and manuals for students and practitioners are also provided. In addition the distribution of university teachers differs in this field. Here the Central, Nordic and Atlantic countries have a longer tradition compared to the Eastern and Southern countries. For example, Sweden and Finland have recently established new chairs in eco-tourism to strengthen the profile for the coming generation of foresters.

There is a correlation between education and working experience: those countries that rate education more positively also rate working experience more positively.

The results show that most of the forest managers gain their knowledge about recreation through their working experience rather than their education. The education, however, is rated as medium-inadequate. This shows a clear deficit throughout Europe.

Usually, the national forestry administration and (regional) forestry management organisations are the main responsible (semi-)governmental bodies for the management of recreation and nature based tourism in forests. All countries were asked to indicate which tasks for which both types of organisations are responsible. The majority responded that the tasks that need to be fulfilled by the different types of organisation are:

- *forestry administration*: planning (81%) and design of the monitoring framework (65%)

- *forest management*: planning (89%), supervising and establishing infrastructure (85%) information (81%), environmental education (77%), management actions for visitor guidance (77%), and monitoring (54%).

There are no clear differences between the different regions, but the results seem to indicate that, for example, the establishment and supervision of infrastructure is more obvious for the Atlantic, Nordic and Central European countries than it is for the Eastern and Mediterranean countries.

7.5 Conflicts

7.5.1 Types of Conflict

In Europe different conflicts arise regarding forest recreation. Within this study five main conflict types concerning recreation in European forests were investigated. Firstly, conflicts may occur between different recreational user groups (e.g. horse-riders – cyclists, cyclists – hikers, joggers – dog walkers) or due to recreational crowding. Secondly, conflicts in forests may also arise between timber production and recreational use (e.g. due to safety issues or aesthetic impacts) or between recreation and other uses like hunting or fishing use. Finally, there may be conflicts between recreation and nature conservation issues.

Across Europe the most important type of recreational conflict occurs between recreation and nature conservation followed by conflicts between recreation and forestry use (Fig. 7.8). Conflicts between recreation and hunting or fishing and between different recreational user groups have a lower level of importance but they are relevant in some countries and regions. Conflicts due to crowding have the lowest level of importance. This conflict type is mainly significant in forests near big cities and in tourism regions.

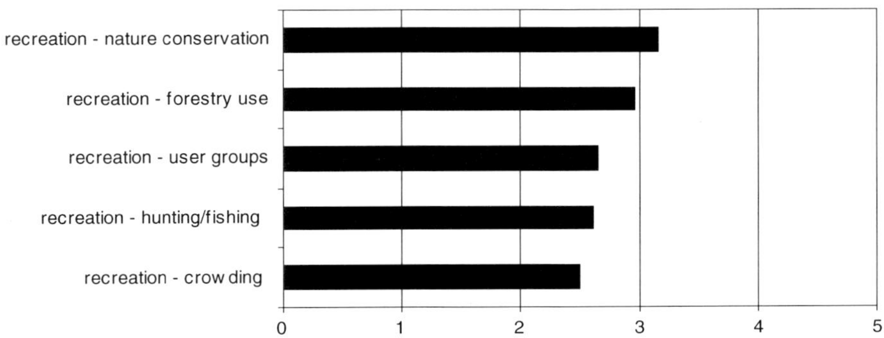

Fig. 7.8 Level of conflict within Europe (1=not important, 5=very important)

If the conflict types are broken down by the different European regions there are significant differences regarding conflicts between recreation and nature conservation and conflicts between different recreational user groups (Fig. 7.9). In terms of recreation and nature conservation, by far the highest level of conflict is found in the Eastern countries. This type of conflict is rated medium-severe in Mediterranean and Central countries, and less severe in Nordic and Atlantic countries. Also, significant differences can be found among conflicts between different recreational user groups. There is a higher level of conflict reported in the Central, Atlantic and Nordic countries and a lower level of conflict in the Eastern and especially in the Mediterranean countries.

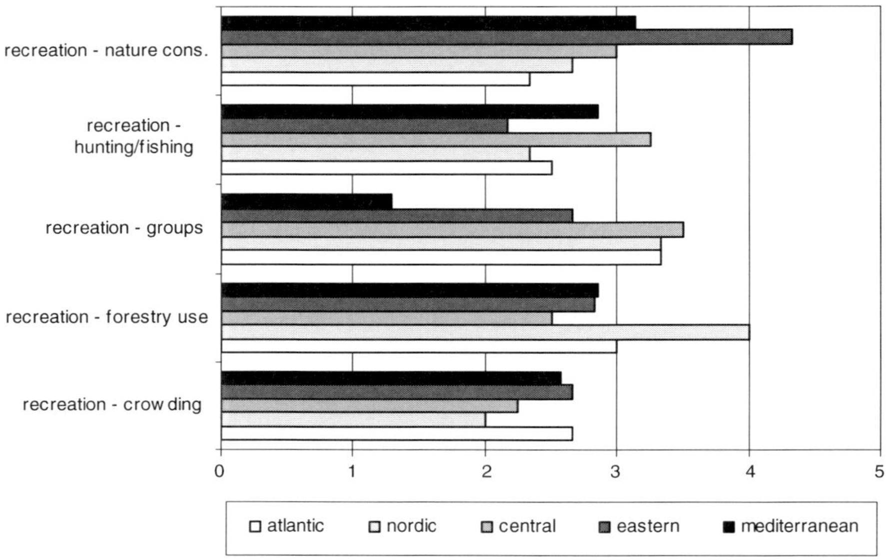

Fig. 7.9 Conflict levels compared between European regions (1=not important, 5=very important) (statistics: nature conservation: $eta^2 = 0.480$ ($p<0.05$), groups: $eta^2 = 0.395$ ($p<0.01$), hunting/fishing, forestry use, crowding: not significant)

Among the remaining types of conflicts, there are less severe differences in the regions, but some aspects are worth mentioning. Conflicts between recreational use and hunting/fishing are more important in the Central and Mediterranean countries. This conflict type is slightly more important in Eastern, Atlantic and Mediterranean countries. In general, we can conclude, that conflicts between recreation and nature conservation are by far most important in the Eastern countries. This fact may relate to the substantial changes that have occurred due to changes of political systems. Large forest areas have been privatised and sensitive areas face new threats. Conflicts between different visitor groups are more important in the more densely populated Central and Atlantic regions. This conflict type is also important in the

Nordic countries. Conflicts here occur mainly near cities and due to motorised recreation, especially snow-mobiling. Conflicts arising from forestry use are most important by far in the Nordic countries. The timber industry is very important in this region and harvesting takes place on a very large scale, diminishing greatly the quality of the forests for recreation. Conflicts regarding crowding are a less important type of conflict - mostly occurring in local or regional recreation "hot spots".

A slightly different picture from the analysis at the regional level is shown at the country level. Figure 7.10 depicts the most important conflict type for each country; in the case of more than one type of conflict being mentioned all highest-scored conflicts are identified.

Fig. 7.10 Most important type of conflict in each country

The conflict between recreation and forestry use is rated as the most important type of conflict by 11 countries, followed by conflicts between different recreational groups (ten countries). Remarkably, all Nordic countries name the conflicts between recreation and forestry use as the most important conflict and all Atlantic countries mentioned conflicts among user groups as most important. Conflicts between recreation and nature conservation and conflicts between recreation and hunting/fishing are rated as the most important conflict in eight countries. Nearly all Eastern countries identify conflicts due to nature conservation issues as the most important conflict. Conflicts due to crowding are only listed as the most important conflict in three Mediterranean countries.

The analysis by country shows some differences compared with the analysis by region. The importance of the different conflict types shifts: conflicts between recreation and forestry use and between different recreational groups are rated more important in this analysis than conflicts between recreation and nature conservation. Generally, however, this analysis by country corresponds to the analyses by region.

7.5.2 Conflicts with Recreational Activities

There are different forest related activities that may cause conflicts concerning the natural environment and nature conservation aspects. Across Europe, conflicts arise most frequently due to picnicking, followed by motorised sports, alpine touring (in countries where this activity is relevant), collecting berries and mushrooms, mountain biking, fishing, campfires, and camping.

In this analysis (Fig. 7.11) countries have not been included for some activities where - due to physical conditions - these activities are not relevant, especially winter sport and climbing activities.

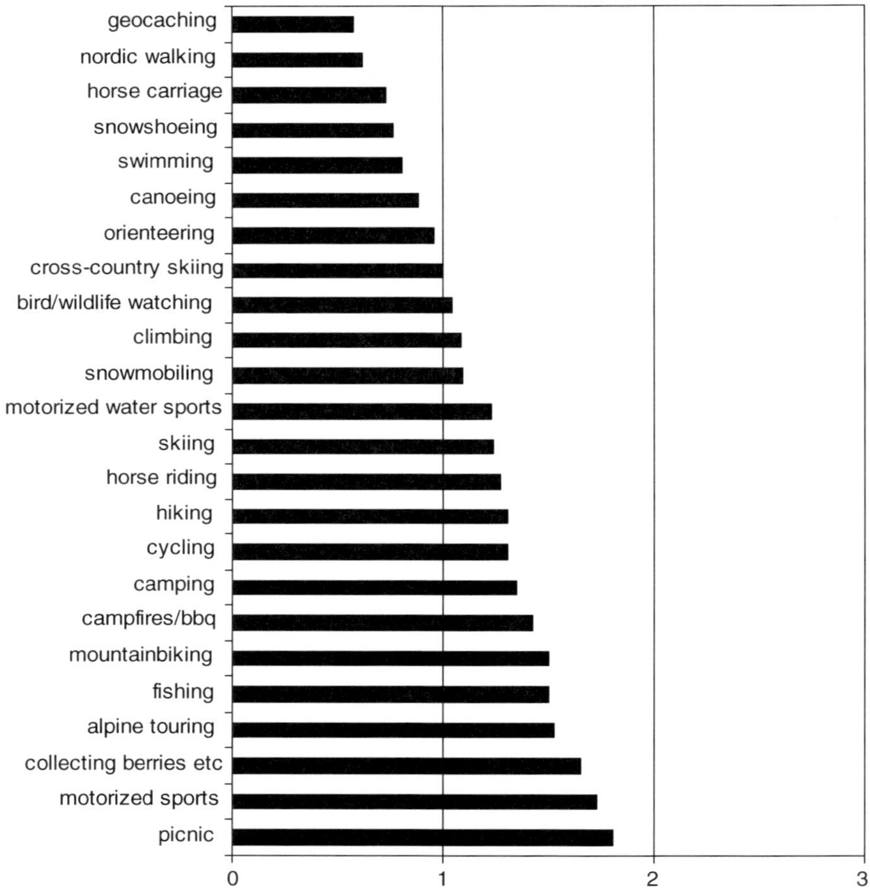

Fig. 7.11 Frequency of conflicts with recreational activities (0=non-existent, 1=rarely, 2=medium, 3=frequently)

The three most conflicting activities are: picnicking, motorised sports, collecting berries and other forest products, alpine touring and fishing. The high level of conflict due to picnicking activities was not anticipated. This result can be explained by the high level of conflict mainly in the Eastern and Mediterranean countries where picnics are big events with large barbecues (with the danger of forest fires). In Cyprus, there are picnic places which can accommodate more than 3,000 persons. In the Central and Atlantic region there also are problems mainly due to rubbish dumping and degradation of picnic sites. Motorised sports are a big problem in the Nordic but also in the Eastern region. In these regions motorised activities are not always regulated. Collecting berries and mushrooms is mainly a problem in the Eastern region and parts of the Northern region where professional berry collectors harvest large amounts of berries, for example 9,900 tonnes in the year 2005 in Poland.

There are significant differences between the European regions for different groups of activities (Fig. 7.12). The activity cluster with the highest level of conflict are activities that use specific locations: bird and wildlife watching, camping, picnicking, campfires and barbecues, and collecting berries and mushrooms. In this activity group no significant differences between the regions are found. However, it these activities appear to cause more conflicts in the Eastern, Central and Atlantic countries. Camping in particular often causes conflicts in Eastern countries, followed by the Atlantic countries; this type of conflict hardly occurs in the other regions. This situation can be explained by the legal situation in different regions: camping is not allowed in forests in the Eastern and Atlantic countries; however, it is a popular activity. An activity that also causes many conflicts in the Atlantic region is bird and wildlife watching.

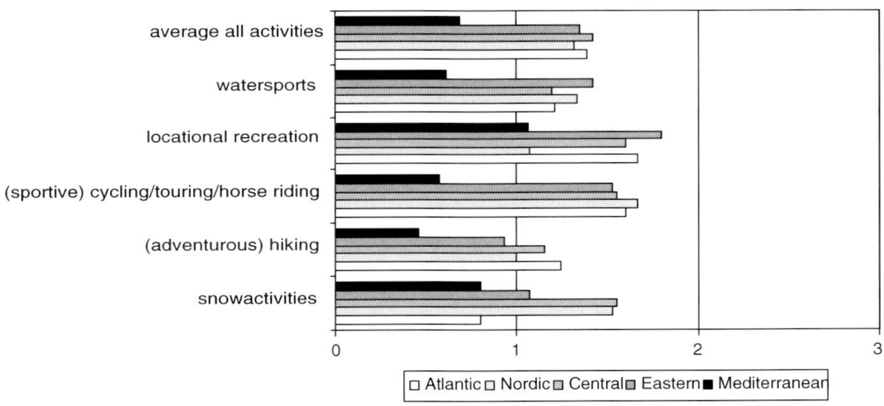

Fig. 7.12 Frequency of conflicts with respect to groups of activities (0=non-existent, 1=rarely, 2=medium, 3=frequently; water sports: eta^2= 0.480 ($p<0.10$), cycling/touring/horse riding: eta^2= 0.549 ($p<0.01$), hiking: eta^2= 0.385 ($p<0.05$), locational recreation and snow activities: not significant; all activities: eta^2= 0.417 ($p<0.05$)

The activity group with the second highest conflict level is cycling/touring/horse riding; here mountain biking, cycling, horse riding, horse carriage driving and motorized sports are included. There are significant differences between the regions:

conflicts due to these activities are most frequent in the Nordic and Atlantic countries followed by the Central and Eastern countries. By far the fewest conflicts arise in the Mediterranean countries. Mountain biking has the highest level of conflict in the Central countries, followed by the Nordic countries and Atlantic countries. The level of conflict is lower in the Eastern countries and especially in the Mediterranean countries. This situation can be explained by the fact that in the Central regions there are many mountainous regions (e.g. Alps, Vosges, and Black Forest) that are very popular for mountain biking because of the terrain. Cycling and horse riding are causing relatively more conflicts in the Atlantic region than in other regions. This may also be a consequence of the legal situation, and the low amount of forest cover in densely populated countries which reduce opportunities for recreation and may lead to more concentration in suitable areas.

There is a medium conflict level for snow activities, which include alpine touring, downhill skiing, cross-country skiing, snowshoeing, and snowmobiling. Although the differences between the regions are not significant, conflicts most frequently arise in the Central and Nordic regions. Particular problems are associated with snowmobiling in the Nordic region and alpine touring and skiing in the Central region. In both regions winter sport activities are very important and also contribute significantly to the regional economies.

Water sports, including fishing, canoeing, swimming and motorized water sports also have a medium conflict level. There are significant differences between the regions. The most frequent conflicts arise in the Eastern and Nordic countries.

Hiking activities, including hiking, Nordic walking, orienteering, geo-caching, and climbing, are causing the fewest conflicts. Hiking, however, still causes some conflicts in the Atlantic countries, whereas this is the case for orienteering in the Nordic countries and for climbing in the Central European countries.

The European comparison shows that the perception of conflicts across Europe is quite different. The experts in the Central and Atlantic countries seem to be more critical in this aspect than their Mediterranean colleagues. These differences may also occur because of the different recreational behaviour and activities in the forest. In the Mediterranean countries many forest related activities do not take place in the same concentrations as in Atlantic or Central countries (e.g. mountain biking) because of the temperatures during the day in summer and because of cultural reasons (like berry picking). Conversely, the tradition of cooking and having a barbecue leads to other conflict types like rubbish dumping (see next chapter) which are much more common activities in the Mediterranean countries.

7.5.3 Environmental Impacts

Several impacts can negatively influence the forest recreational. In this study noise, visual and technical impacts were considered. Impacts can result of noise (traffic noise, aircraft noise, industrial noise, military noise). There can be visual impacts from wind turbines, large clear-cuts, rubbish dumping, fire breaks and visual problems associated with damaged forests and technical impacts. Technical impacts from activities such as road construction, exploitation of natural resources and infrastructure for tourism and recreation can also have a negative effect on recreation.

In European forests, the most frequent impact is rubbish dumping (Fig. 7.13). The next most important impacts are from traffic noise, visual problems of damaged forests, road construction and exploitation of natural resources.

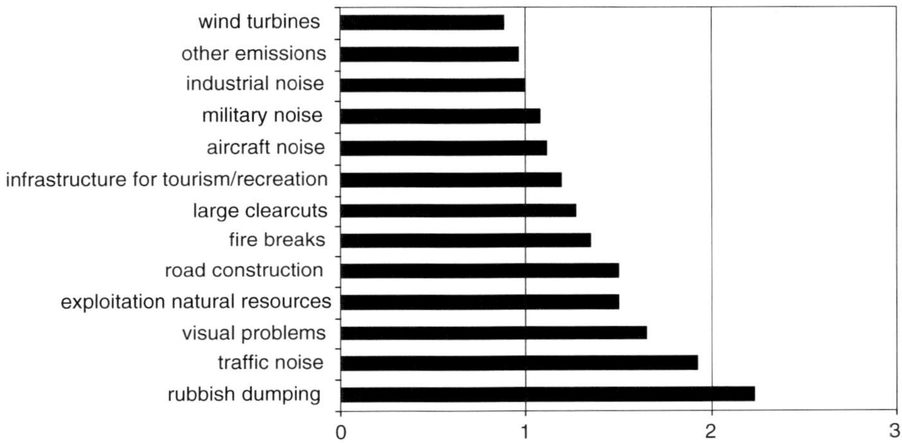

Fig. 7.13 Frequency of environmental conflicts at European level (0=non-existent, 1=rarely, 2=medium, 3=frequently)

Fire breaks, large clear cuts, infrastructure, aircraft noise and military noise are rarely mentioned impacts. Least frequent impacts are those due to industrial noise, other emissions and wind turbines.

Some significant differences can be found between regions (Fig. 7.14). Impacts due to rubbish dumping are most frequent in the Mediterranean countries followed by the Atlantic and Eastern countries. This impact is less frequent in Nordic and Central countries. There are also significant differences regarding visual problems of damaged forests and fire breaks. Most frequently this impact occurs in the Mediterranean countries followed by Eastern countries while being less important in Central, Nordic and Atlantic countries.

In general, we can conclude that rubbish dumping appears to be a major impact – especially in the Mediterranean but also in Atlantic and Eastern countries. Traffic noise is an important impact for forest recreation overall Europe. Mediterranean and most Eastern countries are strongly affected by forest fires due to climate conditions so damaged forests and fire breaks are important impacts in these regions. Wind turbines are locally more discussed, e.g. Germany but are apparently not an important impact at all on European level.

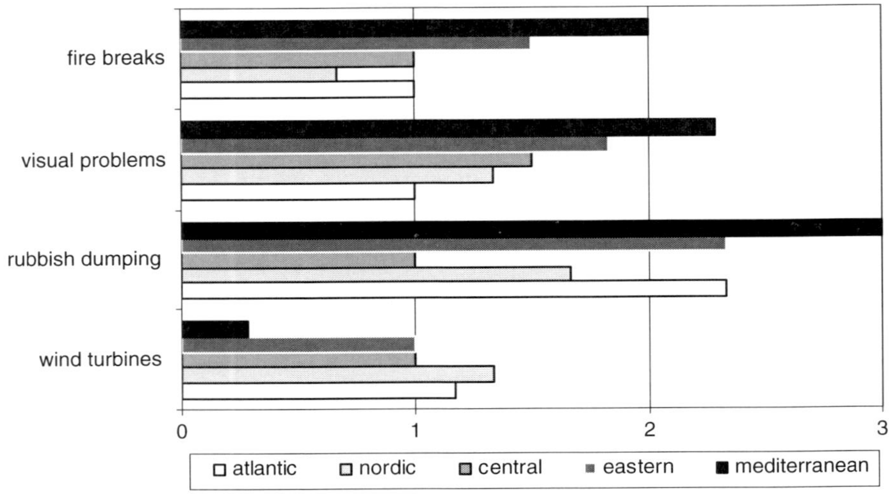

Fig. 7.14 Frequency of examples of environmental conflicts at regional level (0=non-existent, 1=rarely, 2=medium, 3=frequently) (statistics: wind turbines: $eta^2=0.430$ ($p< 0.05$), rubbish dumping: $eta^2=0.606$ ($p< 0.01$), visual problems: $eta^2= 0.501$ ($p< 0.01$), fire breaks: $eta^2=0.412$ ($p<0.05$)

7.6 Planning and Monitoring

Planning, management and monitoring of recreation and tourism in natural elements are three elements of a structured approach to balance demand and supply in a sustainable way. Management tasks have been dealt with in section 7.4. This section deals with planning and monitoring.

7.6.1 Existence of an Independent Planning System

In the majority of the countries there is an independent planning for recreation and tourism in forests. Only in six countries there is no such planning (in Iceland and the Mediterranean countries Bosnia, Cyprus, Croatia, Portugal and Serbia). Figure 7.15 shows the differences between the regions. Only in the Mediterranean region do most countries not have an independent planning system.

If there is independent planning for recreation and nature tourism, in most cases it is optional. Only in the Eastern countries, is it often legally required. Planning systems for recreation and tourism in forests are quite varied throughout Europe and at least three forms can be distinguished:

- planning as part of general land use or landscape planning (e.g. Belgium, Finland, Germany, Switzerland);

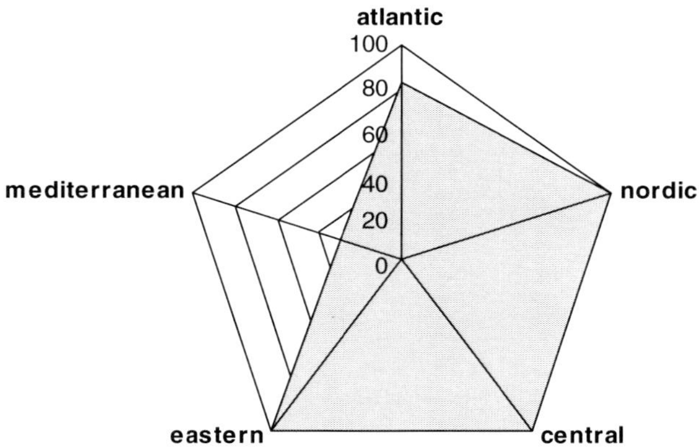

Fig. 7.15 Existence of independent planning for recreation and tourism in forests (% yes; differences between the regions are statistically significant)

- planning integrated in forest inventory or management planning (e.g. France, the Netherlands, Germany, the UK);
- planning in case of special needs of interests (e.g. Norway, Latvia, Austria).

7.6.2 Planning Steps

In general, we can conclude that recreational planning in European forests seems to focus rather at a strategic as opposed to an operational level. This implies that the planning steps that most countries undertake when carrying out recreation related planning includes: definition of vision and goals, public participation in the definition of goals, and inventory of ecological conditions, and to a smaller extent, inventory of social conditions (Fig. 7.16).

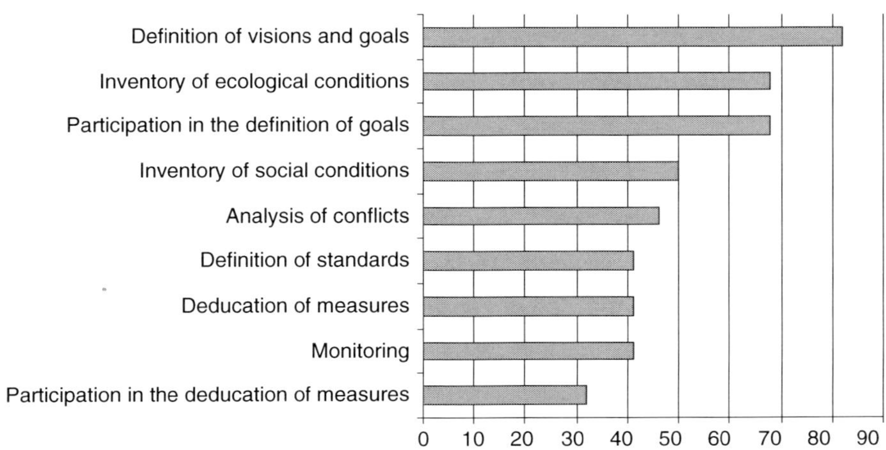

Fig. 7.16 Undertaken steps for planning forest recreation (% yes; N=22, missing countries: Bos, Cro, Cy, Por)

In that respect, it is interesting to observe that for nearly all countries where there is planning for recreation and nature tourism in forests there are handbooks, guidelines or other instructions or frameworks to direct the planning and the management of recreation and tourism in forests. Another source of information is visitor data. Many countries have at least one data source available. These sources vary from data collected by the forestry administration (82%), regional surveys (68%), national surveys (64%) and other forms of data collection being carried out within a particular planning process (64%).

7.6.3 Public Participation in Planning

In most European countries, public participation becomes an issue when visions and goals for forest recreation and nature-based tourism are articulated. Almost all countries (except for the four above mentioned Mediterranean countries, as well as Belgium, Latvia and Serbia) organise some sort of public participation. A variety of social groups can be invited (Table 7.5). Community representatives are the most mentioned group. As land owners act as providers of recreation and tourism on their land, not surprisingly, they take part in public participation regularly. Only mushroom and/or berry collectors are a less common public participation group.

Table 7.5 Involved groups in case participation takes place at European and regional level (N=19)

Groups	Involved (N=19)
Community representatives	95%
Land owners	90%
Land users	90%
Sports clubs	90%
Local residents	84%
Hunting/fishing representatives	84%
Mushroom/berry collectors	16%
Number of participants	5.5

There are small differences between the European regions. Although the Atlantic and Central countries have on average a high urban pressure upon their forests, relatively few land users (Atlantic) or local residents (Central) are involved in public participation than they are in the other regions. As hunting and/or fishing is either an illegal activity or a less common activity in Atlantic forest and nature areas, representatives are less involved in public participation. Mushroom and/or berry picking groups are most frequently involved in Nordic countries. Where the Nordic countries have on average the most social groups involved, the Central countries have the least.

7.6.4 *Monitoring*

Monitoring is an important part of the overall planning process. All countries, except the four above mentioned Mediterranean countries and Latvia, carry out some monitoring activities. Half of them (52%) perform monitoring at irregular intervals or whenever it is considered to be necessary. Another group carries out monitoring less than once every 5 years (14%). A small group conducts monitoring quite regularly, either every 1-2 years (5%) or every 3-5 years (5%). Almost one quarter of the countries (24%) have other monitoring practices. There are no clear observations to be made between regions.

7.6.5 *Overall Evaluation*

Each country expert was asked to evaluate the consideration of recreation and nature tourism in forests, taking into account the national ownership structure, the legal situation, and the education of forest managers. It can be concluded that the majority of the countries (Fig. 7.17) evaluate the overall picture as being "below average". Only eight countries assess the situation of forest based recreation as being good. However, most countries state that the situation is better regarding different regions.

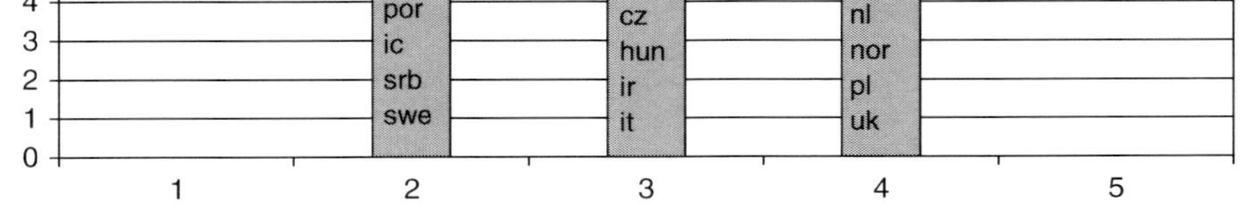

Fig. 7.17 Overall evaluation of the consideration of forest recreation and nature based tourism at a country level (1= unsatisfactory, 5= excellent)

The European score is on average "sufficient" (Fig. 7.18). We can observe that the Atlantic and Eastern countries are most satisfied, whereas the Mediterranean countries are the least satisfied.

Countries where there is an independent planning system rate the consideration of forest recreation and nature tourism higher than countries without an independent planning system (sufficient versus barely satisfactory). The same holds true for countries with public participation and countries lacking public participation.

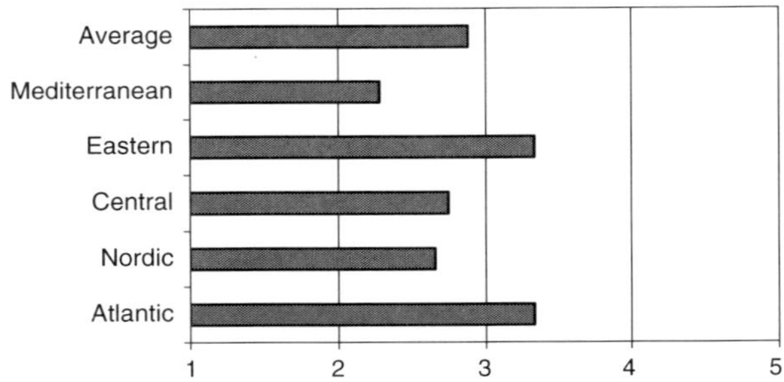

Fig. 7.18 Overall evaluation of the consideration of forest recreation and nature based tourism (1= unsatisfactory, 3=sufficient, 5= excellent; differences between the regions are not statistically significant)

7.7 Recreational Infrastructure

Within this section the basic recreational infrastructure, such as hiking trails and forest roads, and additional infrastructure, such as benches and panorama points, will be assessed. A distinction has been made for urban, rural and remote areas to take into account regional differences all over Europe. It is important to know that both Denmark and the Netherlands consider that they do not have any remote areas; consequently, they are left out of the analysis of remote areas.

7.7.1 Supply of Basic and Additional Infrastructure

From Fig. 7.19 we can see that the basic infrastructure in urban areas is rated higher than for rural areas and that remote areas have the least satisfactory level of basic infrastructure. Overall, we can note that the level of supply varies from "average" to "good". Although the differences between the regions are not statistically significant, it seems that that the Central countries are most positive, followed by the

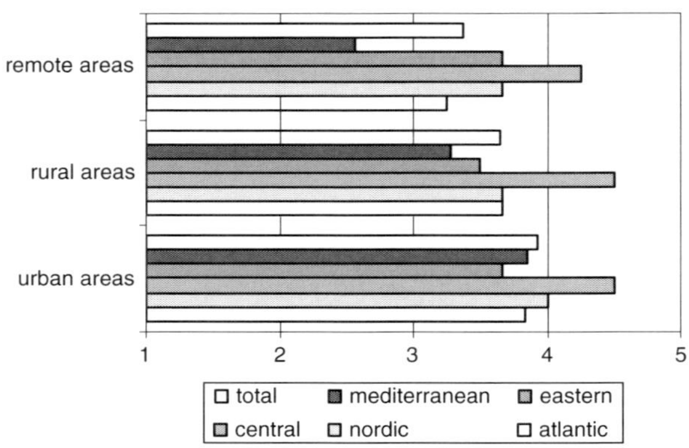

Fig. 7.19 Supply of basic infrastructure in urban, rural and remote areas (1= unsatisfactory, 5=very good) (differences between the regions are not statistically significant)

Nordic and then the Atlantic countries. The Eastern and Mediterranean countries are –on average – least satisfied.

With respect to the supply of additional infrastructure, the Central region rates this the highest, closely followed by the Nordic region and a little less by the Atlantic region (Fig. 7.20). The Mediterranean region is generally relatively unhappy with the present situation.

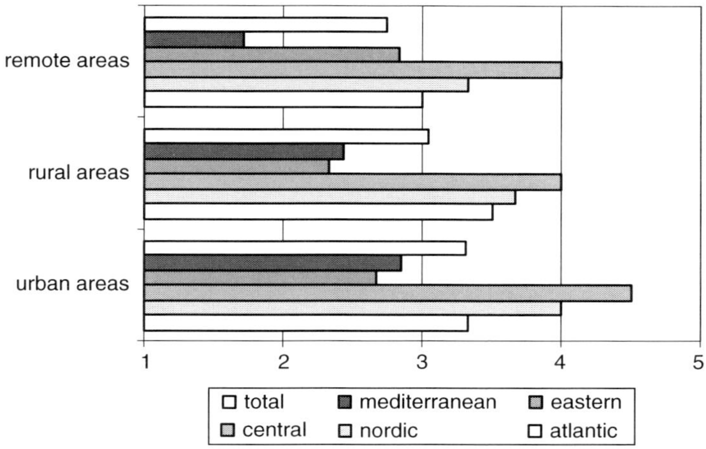

Fig. 7.20 Supply of additional infrastructure in urban, rural and remote areas (1= unsatisfactory, 5=very good) (urban areas: eta^2=0.373 ($p< 0.05$), rural areas: eta^2=0.406 ($p< 0.05$), remote areas: eta^2=0.444 ($p< 0.05$)

7.7.2 Additional Facilities

The availability of additional infrastructure was also evaluated (Fig. 7.21). For almost all facilities urban areas have the highest supply with a large diversity (mean 1.8) compared to rural (mean 1.5) and especially remote areas (mean 1.1). Only for huts, camping grounds, and game preserves can a different pattern be found: huts are most frequently placed in remote areas, camping grounds are equally provided in urban and rural areas and finally, game preserves can be found in any zone. There are no significant differences between the regions.

A second observation is that in urban areas only the availability of benches, play grounds, rubbish bins, picnic areas and nature trails is rated as "medium". Most facilities, except for nature trails and huts, are rarely present in remote areas. This is not necessarily a problem: many experts indicated that the more remote a forest and nature area becomes, the more deliberate the policy is on restricting the number of facilities. Furthermore the demand or need for them is not as high in rural and remote areas.

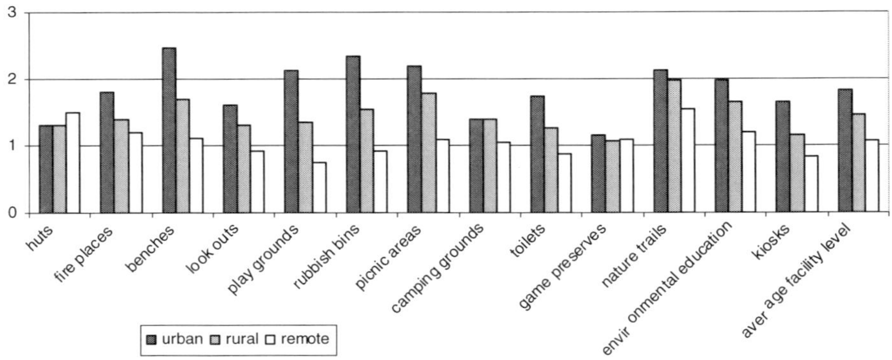

Fig. 7.21 Additional infrastructure facilities in urban, rural and remote areas across Europe (0= non-existent, 1=rarely, 2=medium, 3=frequently) (differences between area types are mostly not statistically significant)

7.7.3 *Supply of Infrastructure: Over-equipped?*

In the past, experts from Central Europe rated many forests as being over-equipped for experiencing nature (Ammer 1983). It appears, however, that most countries do not feel this to be the case (Fig. 7.22). In this respect it is interesting to observe that the Central European countries, especially France, Germany and Switzerland, as well as the Netherlands and Greece are the only countries who perceive their forests to be over-equipped. This concerns forests in urban areas in particular. This may be associated with increased urbanity, and extensive provision of facilities and activities in the forest. These increases may produce a greater need for the user to experience untouched nature in forests and other natural areas.

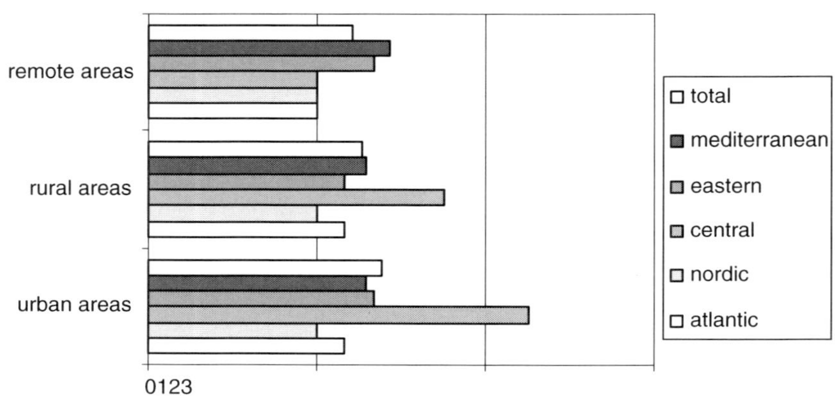

Fig. 7.22 Opinions on over-equipment of forests for experiencing nature in urban, rural and remote areas across Europe (1=disagree, 3=agree) (urban areas: eta^2=0.313 ($p < 0.05$), rural and remote areas not significant)

7.7.4 *Quality of Infrastructure*

The quality of both the basic and additional infrastructure is very important for recreation and tourism. Quality includes good original design, regular innovations, maintenance, checking of safety features, and adaptation of design over time. The country experts were on average the most satisfied about the quality of infrastructure in urban forests and the least about remote areas (Fig. 7.23). It can also be concluded that the Mediterranean countries are least satisfied with the quality of infrastructure, no matter what area type. The Nordic, Atlantic, and Central European countries are most satisfied, although none of the regions rates the infrastructure in as "excellent". The Eastern countries hold an intermediate position between the other regions.

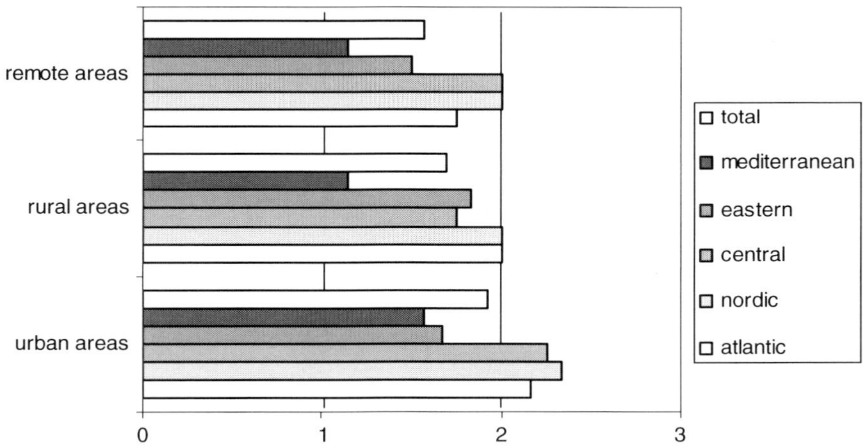

Fig. 7.23 Quality of infrastructure in urban, rural and remote areas across Europe (1=unsatisfactory, 3=excellent) (remote areas: $eta^2=0.450$ ($p< 0.05$), urban and rural areas not significant)

7.8 Visitor Management

There is a large variety of actions for visitor management in forests. In this study actions for visitor management in forests are divided into three groups: restrictive, soft and facilitating management actions:

- *Intensive management actions* influence the access or behaviour of visitors directly. In this study implementing rules and regulations, prohibitions, entrance fees, marking protected areas and taboo zones, and limiting visitors' entrance are included in this group.
- *Soft management actions* influence visitor behaviour indirectly or appeal to the responsibility of the visitors. In this group improving areas with new attractive

infrastructure, new infrastructure for ecological education, signposting for visitors to behave in a responsible way and guiding visitors with a ranger system are examples.

- *Facilitating management actions* are listed as separate group because they focus explicitly on providing training and education to visitors. These actions can enhance visitors' awareness and understanding of forest ecosystems. Here examples include guiding visitors with thematic tours and providing special training for user groups and for commercial providers.

7.8.1 Use of Different Types of Management Actions

Figure 7.24 shows how often management actions are used in European forests in urban, rural and remote areas. It can be concluded that soft actions are the most common, followed closely by restrictive management actions; facilitating actions being least used. Secondly, in urban areas most management measures are implemented, whereas in remote areas the fewest management actions are implemented.

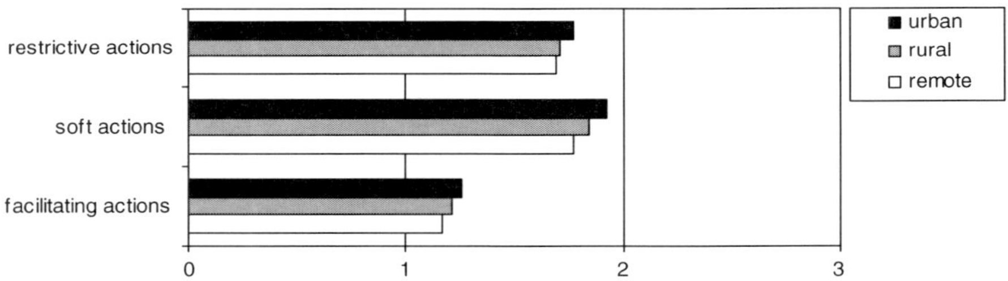

Fig. 7.24 Restrictive, soft and facilitating management actions in urban, rural and remote areas across Europe (0= non-existent, 1=rarely, 2=medium, 3=frequently)

If we focus more in detail on the character of the management action, we can see that those most frequently used are rules and regulations and prohibitions, which both belong to the category of restrictive actions (Fig. 7.25). The next most frequently used actions belong to the group of soft management actions: improving areas with attractive infrastructure and signposting for visitors to behave in a responsible manner. They are used more often in an urban setting. Marking protected areas and taboo zones, erecting barriers and fences, new infrastructure for ecological education and guiding by thematic tours are used occasionally. The most rarely used management actions in are restricting visitors' entrance, demanding an entrance fee, guiding by a ranger system, training special user groups and training commercial providers.

It should be noted that the more urban the location of the forest, the more frequently are most management actions used. Exceptions are marking protected areas and taboo zones, demanding an entrance fee, and guiding visitors by thematic

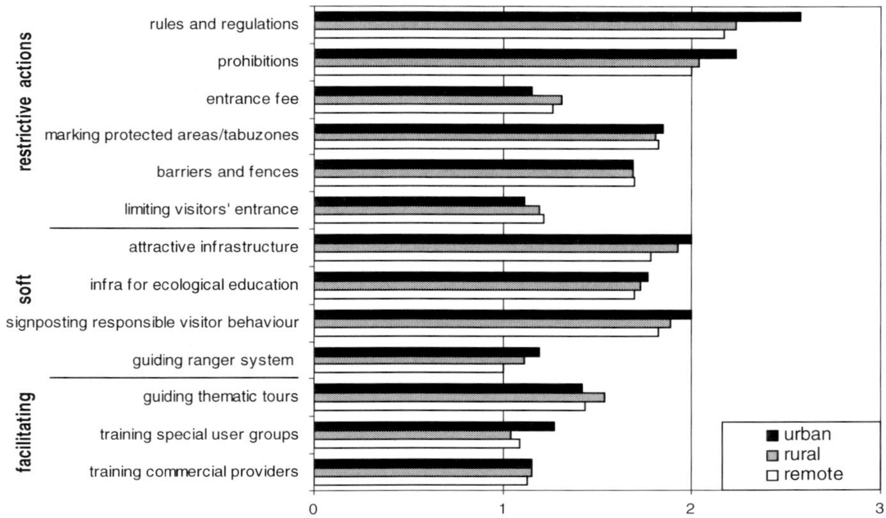

Fig. 7.25 Management actions in urban, rural and remote areas across Europe (0= non-existent, 1=rarely, 2=medium, 3=frequently)

tours. These actions are used relatively more often in less centrally located forests. These results seem plausible because higher visitor density requires more frequent management actions. Some actions, however, are more suitable for remote areas: protected areas tend to be located in remote areas, visitor fees and guiding activities are used more commonly in tourism and day trip destinations than in recreation "hot spots".

The analysis of regions shows no big differences (Fig. 7.26). In all regions except the Mediterranean region, soft management actions are used most frequently; in the Mediterranean region restrictive actions are used most frequently. In all regions facilitating actions are used least frequently. Soft management actions are used most frequently in the Eastern countries. Restrictive actions are used most frequently in the Eastern, Mediterranean and Nordic countries. Facilitating actions are used least frequently in the Mediterranean and Central region, slightly more frequently in the other regions.

In the regional analysis of single management actions it becomes obvious that demanding an entrance fee and limiting visitors' entrance are management actions predominately used in Eastern countries (mean 1.6) and Mediterranean countries (mean 1.5) and far less so in the other regions. Rules and regulations are used most frequently in the Nordic countries (mean 2.7) and Eastern countries (mean 2.6); they are used less in the other regions. Barriers and fences are used most frequently in the Atlantic region (mean 2.0), the Mediterranean (mean 1.8) and Eastern region (mean 1.7).

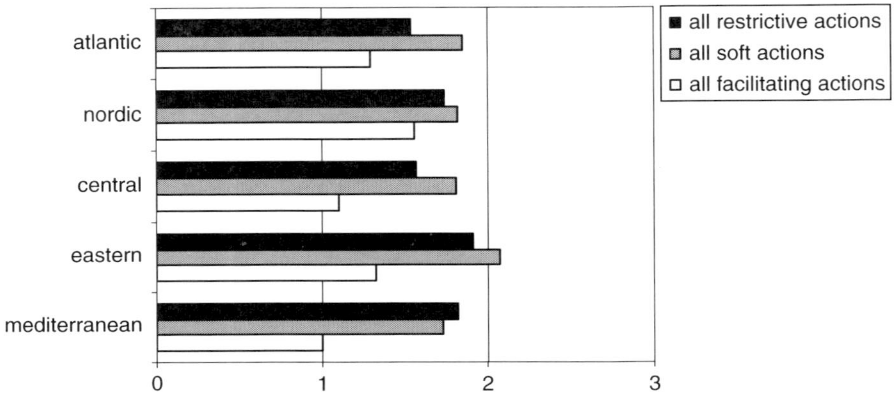

Fig. 7.26 Management actions summarized in European regions (0= non-existent, 1=rarely, 2=medium, 3=frequently)

7.8.2 *Acceptance by the Public*

The experts of each country also rated the acceptance by the public of the various management actions (Fig. 7.27). Generally, the most accepted management actions are soft actions, followed by facilitating actions. The least accepted management actions are restrictive actions. Across Europe entrance fees, barriers and fences, limiting visitors' entrance and prohibitions are the least accepted management actions. Marking protected areas and taboo zones is the best accepted management action of all intensive management actions. Improving areas with new attractive infrastructure and new infrastructure for ecological education are the most accepted management actions.

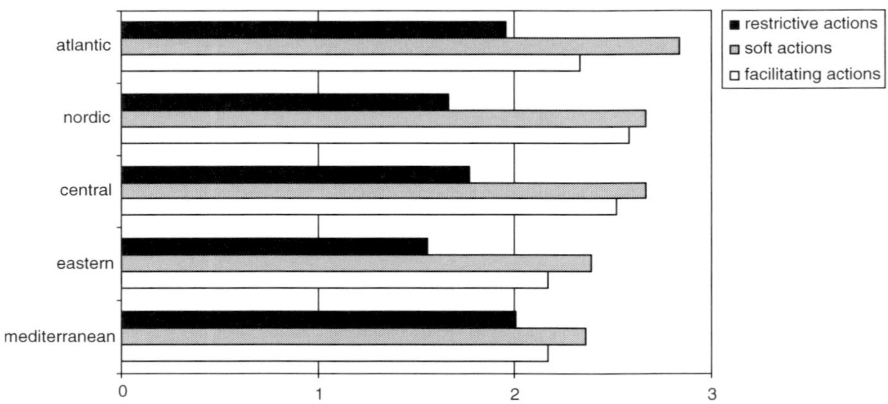

Fig. 7.27 Acceptance of management actions in Europe and the regions (1=rarely, 2=medium, 3=high) (differences between regions are not statistically significant)

There are only small differences between the regions: the highest acceptance of soft management actions can be found in the Atlantic countries, the greatest acceptance of restrictive actions in the Mediterranean countries.

If the acceptance of the different management actions is analysed in more detail, there are statistically significant differences for some actions:

- *Rules and regulations* have a medium level of acceptance across Europe as a whole, though there are significant differences between the regions. There is the highest acceptance in the Atlantic countries, followed by the Central and Nordic countries. Acceptance is lower in the Mediterranean and Eastern countries.
- *Entrance fees* have generally a low level of acceptance but there are significant differences between the regions. The best acceptance by far can be found in the Mediterranean countries. The acceptance by the public is average In the Eastern and Atlantic countries. A low acceptance by the public is found in the Central and Nordic countries.
- *Erecting barriers and fences* are generally the least accepted action for visitor management in forests. However, there are also significant differences between the regions. This action has a higher degree of acceptance in the Mediterranean and Atlantic countries. The acceptance is lower by far in the other regions: Eastern, Nordic and Central countries.
- *Guiding visitors* by a ranger system is one of the best accepted actions for visitor management in Europe, though there are also differences between the regions. There is a high degree of acceptance in the Central and Atlantic countries, followed by the Mediterranean, Nordic and Eastern countries.
- *Providing training for commercial providers* is a fairly well-accepted management action with significant differences between regions. A greater acceptance is found in the Nordic and Eastern countries. The degree of acceptance is lower in the Atlantic and Central and Mediterranean countries.

We can conclude that there are some striking, probably cultural differences between the regions. Rules and regulations seem to be more accepted in regions that are densely populated like the Atlantic and Central countries. It is striking that the acceptance of entrance fees and barriers and fences in forests are generally low in countries and regions with relatively free access rights or where only minor restrictions on recreational use may be implemented in forests. In terms of facilitating management actions, there appears to be more potential: these management actions are highly accepted by the public, but only rarely used in all regions.

7.9 Conclusion, Problems and Challenges

We can see from this chapter that across Europe there are significant differences in forest cover, forest ownership, and population density. This situation leads to different problems and so different solutions are needed. This is why the study uses a regional approach differentiating the five European forest regions with similar cultural, historical, traditional and climatic conditions influencing the activities and recreational conflicts in forests. Similarities regarding planning and management approaches can also be found. Figure 7.28 shows the characteristics of each region.

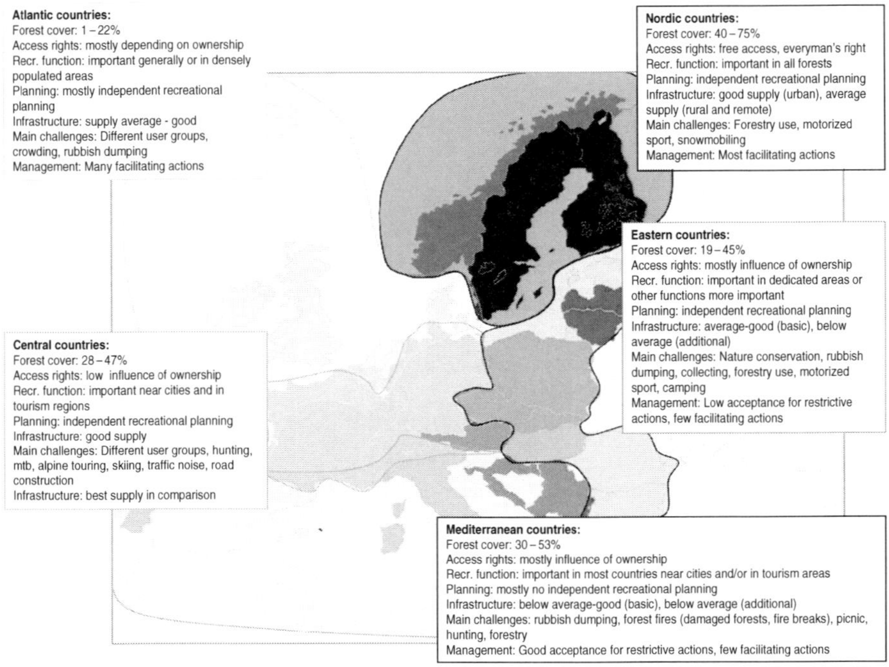

Fig. 7.28 Regional comparison across Europe

We can conclude that the degree of urbanisation in combination with the amount of forest cover determines to a large extent the visitor pressure upon the existing forests and consequently the importance of recreation. This becomes especially apparent in the Atlantic region and to a smaller extent in the Central region.

All over Europe there are laws or regulations affecting recreational use in forests. However, there are large differences in questions of recreational access to the forests. In many countries in the Nordic and Central region as well as in Iceland and the United Kingdom (Atlantic region), visitors have a fairly free access to all natural areas. In many other countries, e.g. in the Atlantic and most of the Eastern and Mediterranean region, state or publicly owned forests permit free access, whereas privately owned forests have limited access.

A challenge for the planning and management of forest recreation and nature based tourism is to improve the access especially around cities and in tourism regions. An example here can be Scotland where the Land Reform (Scotland)

Act 2003 grants access to the countryside for recreational and educational purposes. In so doing it also specifies the responsibilities of both land owners and visitors.

However, there is also the opposite development. In countries with a comparatively free access right areas can closed to recreational use. Austria can be given as example. Mainly in the alpine area deer protection zones can be established in areas where there is mainly private ownership. These areas are mainly used for hunting purposes and recreational access is forbidden.

We can conclude that across Europe that there is a shift of focus of forest functions. Recreation has become very important in many countries, especially around urban centres. For many different recreational activities, including new ones, forests are a very valuable recreational setting. The significance of the recreational function in forests varies across the regions. It is striking that in the Nordic region, where there is the best access to the forests, recreation is also considered important in all forests. In the most urbanised Atlantic countries, such as Belgium, the Netherlands and the United Kingdom, recreation is the most important forest function in many areas. In the Central countries with comparatively free access rights recreation is also considered locally an important function. It is to be expected that in the Eastern countries the importance of recreation will increase according to the further development of their economies and societies. This may increase the challenges and the need for planning and management in these countries.

With respect to the increasing value of recreation, it is important to understand that "knowledge" is an essential requirement for the consideration of forest recreation and nature based tourism in planning, management and monitoring. Although the majority of the European countries recognise this, at the same time most countries are not very satisfied with the present level of knowledge to be found amongst forest managers. Knowledge can be acquired through both working experience and education. In general, the relevance of working experience was rated higher than the relevance of education. In the field of education the deficiencies in the Mediterranean countries are obvious. Here is one of their main challenges. There is a correlation between education and working experience: those countries that rate education more positively, also rate working experience more positively. The results show that forest managers gain their knowledge about recreation as a result of their working experience rather in their education. Education, however, is rated as medium-inadequate. This shows a clear deficit on a European level. Forest managers' knowledge should be developed at two levels: the forest administration of a country, which is responsible for strategic planning and development of a monitoring framework has to acquire different more abstract knowledge than the forest management group, who are responsible for the more operational aspects of planning and management at a site level. In their daily practices, forest managers are confronted with many conflicts that are related to or caused by recreation. The complexity of these problems has increased and will continue to increase due to the growing importance of recreation as well as the growing visitor pressure.

In this study, conflict types, conflicts with respect to specific activities and impacts on forest recreation were analysed. The conflict type with the highest conflict level in European forests arises between recreation and nature conservation and between recreation and forestry use. Activities causing conflicts most frequently with regards to nature conservation in Europe are picnicking, motorised sports and alpine touring. The most frequent environmental impacts are rubbish dumping, traffic noise and visual problems of damaged forests. However, there are very different challenges in the different regions.

In the Atlantic region most conflicts arise between different user groups, due to recreational crowding, and rubbish dumping. A challenge for recreation planning and management would be to use spatial zoning techniques, to provide better access to the forests as well as to establish (new) forests and natural recreation areas around urban centres in order to have more recreation opportunities available.

In the Eastern countries most conflicts arise between recreation and nature conservation and forestry use. This fact may relate to the substantial changes that have occurred because of the change of the political system. Large forest areas have been privatised and sensitive areas face new threats. There are also recreational activities leading to conflicts: collecting activities, motorised sport and camping, rubbish dumping also being a major conflict. Here, the challenge for recreation planning and management is to manage and channel these activities and to raise awareness with user groups.

In the Nordic region the most important conflicts arise between recreation and forestry use. In this region, the timber industry is very important and harvesting takes place on a very large scale diminishing the quality of the forest for recreation. In this region the challenge will be to develop new nature based tourism offers which capitalise landscape values.

In the central region most conflicts arise between different user groups and between recreation and hunting. There is also a trend for sport activities like alpine touring, climbing and snowshoeing to cause conflicts. A challenge will be to develop solutions for these conflicts taking into account both the need to provide attractive offers and the protection of the fragile ecosystems where these activities take place.

In the Mediterranean region many conflicts arise between recreation and forestry use and hunting respectively. However, most conflicts arise due to forest fires. Here a challenge is to develop efficient strategies and raise awareness to fight and prevent fires. In the different regions different good practice examples have been developed to tackle the above named conflicts (see Chapter 8).

Good, independent planning systems can assist the forest manager to cope with all these conflicts. In the majority of the European countries, some sort of independent planning system has been developed. Only in most Mediterranean countries is this insufficiently organised. Planning systems, however, are quite varied: some countries deal with recreation and tourism as part of integrated land use or landscape planning, others focus at a site level by means of a forest inventory or management planning system, and again others only plan occasionally in case of special needs or interest. We can observe that across Europe planning systems are diverse and not equally well developed. Moreover, the majority are not satisfied with the consideration of forest based recreation and nature tourism in the forests of their country, taking into account the national ownership structure, the legal situation and the education of forest managers.

The most important conclusion we can draw is that in terms of planning, management and monitoring many improvements can be established in Europe. In many countries there is visitor data available, however, it is not always collected systematically and comparable throughout the different forest areas/regions. A challenge for the planning and management of forest based recreation will be to develop and establish visitor monitoring systems and strive for a harmonization of existing systems (Sievänen et al. 2008). Recreation planning in European forests seems to focus at a strategic level rather than an operational level. A challenge will be to establish an operational level of planning. It also has to be noted that recreational planning has to take the different requirements in urban and rural areas more into account. Different good practice examples for the planning of forest recreation and nature based tourism have been compiled (see Chapter 8).

This chapter also evaluated the supply of basic and additional infrastructure in urban, rural and remote areas. It can be concluded that on average the supply of basic infrastructure, which consists of hiking trails and forest roads, is considered to be average to good. The level of satisfaction with the supply of additional infrastructure, which consists of for example, huts, benches, play grounds, rubbish bins, picnic areas and environmental education facilities, is somewhat lower. The availability of infrastructure is different for different types of areas: urban areas have most facilities, whereas remote areas have fewest facilities. This is not necessarily problematic: many experts indicate that the more remote a forest and nature area gets, the more deliberate the recreation policy is on a restricted number of facilities. This relates self-evidently to the desired wilderness experience for visitors. It is stated that not every type of area should have an equal level of infrastructure provision. Moreover, several countries are of the opinion that there is a risk of over-equipment for experiencing nature. Countries with a relatively high urban pressure and an already extensive supply of facilities and activities in the forests, situated in the Atlantic and Central regions, express this concern. A challenge for recreation planning and management would be to diversify the supply for different experiential user groups, since some people prefer more highly equipped areas in which nature can be easily consumed and other people prefer less equipped areas because they are looking for solitude and unspoilt nature

In the densely populated regions there is often a very dense network of forest roads. A challenge for recreation planning and management would also be to conserve old narrow paths and trails which are often more attractive for recreation seekers and tourists. In Austria, in some alpine areas many new forest roads are built, greatly diminishing the recreation experience. By contrast, in the Eastern region, infrastructure is in many cases lacking. A challenge for recreation planning and management will be to establish recreational infrastructure to manage and direct visitor flows and activities in recreation hot spots with the problem of degradation and in sites with conflicts due to nature conservation.

In many countries infrastructure was established some years ago. A challenge will be to modernise this infrastructure and create attractive offers for recreation and nature based tourism. Also the use of new technologies can be reasonable to create offers using new technologies (e.g. GPS in the environmental education).There are different demands on recreational infrastructure that make a regional adaptation necessary. Different good practice examples of recreational infrastructure have been compiled (see Chapter 8).

A final topic that is dealt with in this chapter is the implementation of management actions in urban, rural and remote forests. It can be concluded that soft actions are most commonly implemented, followed closely by restrictive management actions; facilitating actions are the least used. Secondly, in urban areas, the most management measures are implemented, whereas in remote areas the fewest management actions are implemented. An exception has to be made for marking protected areas and taboo zones, demanding an entrance fee, and guiding visitors by thematic tours. These actions are used relatively more often in less centrally located forests. These results seem plausible because higher visitor density requests more frequent management actions. Some actions, however, are more suitable for remote areas: protected areas tend to be located in remote areas, visitor fees and guiding activities are used more at tourism and day trip destinations than in recreation "hot spots". Regarding the public acceptance of management measures, we can conclude that the frequency of use of actions for visitor management in forests and the acceptance by the public correspond in many cases. Thus, improving areas with new attractive infrastructure and signposting for visitors to behave in a responsible way are frequently used and well-accepted management actions. Entrance fees and limiting visitors' entrance have a low degree of acceptance and are rarely used. It is obvious that across Europe soft management actions have the best degree of acceptance by the public. It has to be noted, however, that soft management actions are not in all cases applicable, sometimes restrictive actions are necessary, especially if an area is heavily used and in danger of being degraded.

Regarding facilitating management actions, there is potential: these management actions are highly accepted by the public, but only rarely used in all regions. However, for these actions trained personal on site is needed. A challenge for the management of recreation and nature based tourism will be to reduce differences across Europe, also considering a development of nature based tourism offers. A further challenge would be to use more facilitating actions, also using new technologies (e.g. GPS or pocket PCs to guide visitors and to provide information).

All across Europe there are differences regarding forest recreation and nature based tourism arising partly due to historical, traditional and climatic conditions, while culture and education also play a significant role. A common challenge across Europe will be to improve the knowledge of recreation in the forest managers' education. In general no Europe-wide solutions can be presented but tailor-made solutions for the requirements of different regions are necessary. The following chapter illustrates good practice and tailor made solutions in recreation planning and management in European forests.

Chapter 8
Good Practice in European Recreation Planning and Management

8.1 Introduction

Berit C. Kaae

8.1.1 Forest Recreation and Nature Based Tourism – a Dynamic Phenomenon

This chapter describes a number of good practice examples in outdoor recreation and nature-based tourism collected from different European countries. The examples provide trans-national inspiration on how to solve some of the problems and challenges identified in the previous chapters. Furthermore, it includes a range of good examples of initiatives that enhance forest recreation experiences. The examples represent a selection from over 100 good practices examples submitted by participants of the COST action E33 "Forest Recreation and Nature based tourism", working group "recreation planning and management". Examples have been submitted by more than 20 countries across Europe.

Forest recreation and nature-based tourism is an evolving and changing phenomenon reflecting the general trends in society and globalization processes. In many countries, the role of the forest as primarily for timber production is losing ground to other societal needs, especially the increasing interest in outdoor recreation and nature-based tourism. Forest recreation and nature-based tourism both include a range of traditional activities now pursued for leisure (e.g. hunting, fishing, horse riding) and a constantly emerging range of new recreational activities combining current leisure trends with technological innovations (e.g. kite wing, mountain biking or blowkarting).

B.C. Kaae (✉)
Dep. of Urban and Landscape Studies, Danish Centre for Forest, Landscape and Planning, KVL, Rolighedsvej 23, 1958 Frederiksberg C., Denmark
e-mail: bck@life.ku.dk

U. Pröbstl et al. (eds.), *Management of Recreation and Nature Based Tourism in European Forests*, DOI 10.1007/978-3-642-03145-8_8,

Within a few generations, many European countries have moved from economies based mainly on primary production through industrialization and secondary production to the service society and now are trending towards the experience economy (Pine and James 1999) are emerging. Urbanization, the information society and technological innovations have also driven significant changes and opportunities. Many of the forests have essentially remained the same throughout these societal changes but the way people look at and interact with forest settings has changed and poses new challenges to forest planners and managers. With increasing urbanization, the severing of ties with the countryside and nature in many people's lives and technological innovations, forestry and other primary production sectors are becoming less and less connected to the everyday life of most people. Consequently, educational and interpretative efforts gain importance in forest recreation, in order to foster an understanding of the values of and roles played by forests and other natural environments.

8.1.2 New Challenges for Management

Today, as noted in the earlier chapters, forest planners and managers are facing many new challenges in meeting the recreational demands of an increasingly diverse range of forest recreation and nature-based tourism activities. How can these diverse activities best be integrated, planned and managed in the forest settings in balance with other forest functions? How can conflicts be avoided and how can the forest experiences be enhanced and made accessible to all, especially to users of different abilities? No quick and easy solutions exist but through our collection of good practice examples in forest recreation and nature-based tourism, we hope to contribute to a trans-national exchange of good ideas and possible solutions to help forest planners and managers meet some of the many challenges they face.

Forest-planners and managers, as well as recreational visitors, may gain inspiration from the range of good practice examples presented in this chapter. Good practices can be aimed at many levels and be of use by different actors. As shown in Fig. 8.1, the key interactions are naturally between the forest planners and managers, the recreationists, and the sites and facilities in the forest.

8.1.3 Good Practise Examples in Europe

For forest planners and managers, recreation is often only one of many responsibilities in the forest and they may have different degrees of knowledge about recreation and nature-based tourism. Good practice examples regarding planning and management strategies show how forest planners and managers can improve the planning and management of recreation sites and facilities to meet the needs of recreationists and tourists. The focus is on different approaches to recreation planning, some typical planning processes, existing planning tools and guidelines as well as on management concepts, especially certification and auditing schemes and risk management.

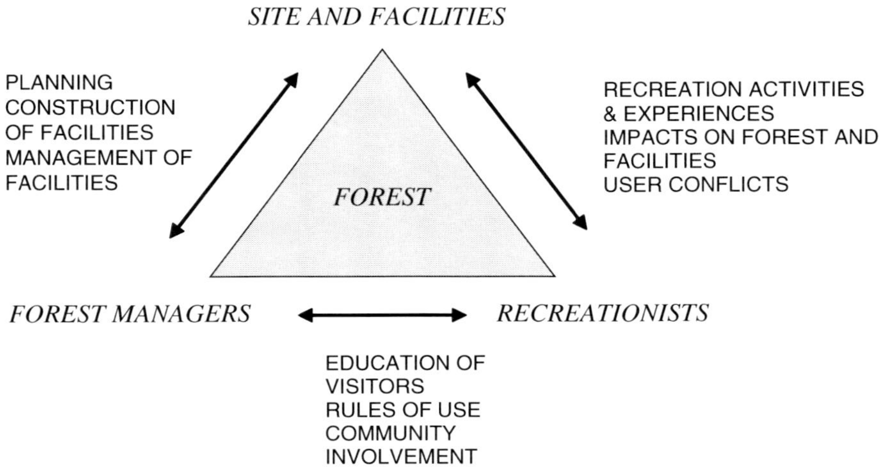

Fig. 8.1 Interactions between forest managers, recreationists and site and facilities

In the main section in this chapter we present a number of examples of detailed management actions and specific solutions for recreation sites in forests. The previous chapters described the characteristics of forest recreation and nature-based tourism across the different European regions. Now we provide some solutions which fit the contexts of these regions. The first part of the section shows positive examples and different ways how to solve these conflicts in different European regions. The second part deals with the supply for recreation in forests by providing examples of good forest design and design of recreational infrastructure to enhance the recreational experience. Some recreation activities need more facilities than others and some are more linked to specific settings than others. The sites and facilities have also to be balanced against the risk of the forest becoming over-equipped or developed, depending on whether it is in an urban, rural or remote area, as well as on the local demand and visitor types. The third part deals with how to provide unique experiences for visitors in forests. Here original and innovative examples from different European countries show new and innovative approaches in the fields of adventure, imagination, health, environmental education and spirituality. Sharing good examples between countries can contribute to inspiration and to the use of alternative approaches in planning, design and management of forest recreation and nature-based tourism.

8.1.4 Regional Differences in Problems and Solutions

The regional differences in forest cover, forest ownership, recreational traditions and population density across Europe (summarised in Chapter 7) naturally lead to regionally different problems and varying approaches and solutions. While some problems may be shared across all regions (e.g. impact of traffic noise) some are specific to certain regions (e.g. fire risk in the dry southern regions) and some are specific to certain sub-regions (e.g. avalanche problems in mountainous areas). Consequently, not all the good practice examples may be relevant for all regions. Furthermore, national differences in culture, planning and management approaches may affect the transferability of some good practices to other countries. Just because solving problems in one country works well does not mean that it will necessarily do so in another context.

This recognition of regional variation is not only a matter of differences in the planning and management "culture" of different authorities responsible for forest recreation and nature based tourism. It is also the acceptance by the public of different actions for visitor management in forests, which may vary regionally as does the willingness to accept certain rules and regulations (see Chapter 7).

8.2 Planning Forest Recreation

Ulrike Pröbstl,[1] Veronika Wirth and Berit C. Kaae

With contributions from: Jan Blok, Kevin Collins, Art McCormack, Birgit Elands, Joel Erkkonen, Tessa Hegetschweiler, Savvas Kazafaniotes, Gudrun von Langehove, Barbara Mariotti, George Pattichis, Jon Geir Petursson, Pieter Roovers, Dan Rydberg, Odd Inge Vistad, Dijana Vuletic, Roger Worthington.

8.2.1 Planning Types

Compared to the North American planning culture which is dominated by the development of frameworks (such as the Recreation Opportunity Spectrum) European recreation planning is mainly characterized by tailor made solutions adapted to specific local problems. Furthermore, recreation planning involves different tasks:

1. *In combination with regional development: Consideration of recreation in forests at a national or regional level*:
Here the main aim is to ensure that there is a sufficient amount of forest, accessible to the majority of people, with a network of main paths related to the proximity of housing areas. These plans are mostly appropriate for the surroundings of cities and for densely populated areas. It is often the aim to protect these forests and to dedicate them as "recreational forests". Recreational forests can be found as part of the regional plan e.g. around Munich and in Augsburg in Germany.

[1]University of Natural Resources and Applied Life Sciences, Institute of Landscape Development, Recreation and Conservation Planning, Peter-Jordan-Strasse 82, 1190 Vienna, e-mail: ulrike.proebstl@boku.ac.at

2. *In combination with conservation planning*:
A second field for large-scale recreation planning is in combination with nature conservation planning and the planning of protected areas. In the case of the development of a biosphere reserve in Münsingen, Germany, nature conservation goals were combined with recreational planning.

In Poland the enlargement of the Białowieza Primeval Forest with the last European bison is another planning example which combines nature conservation and recreation aims in order to enhance the regional development of an area close to the Belarussian border. The enlargement of the bison area includes additional recreational and interpretative facilities as well as better access for recreation.

Box 1: Recreation planning in context with conservation planning

Sveaskog's Ecoparks combine management of nature conservation and recreation at a landscape level. An ecopark is a large (>1,000 ha) continuous forest landscape with high natural values that also serves as arena for local nature tourism operators. Today 36 ecoparks have been inaugurated. Ecoparks play an important part in the implementation of Sveaskog's environmental policy. Ecoparks are developed as prioritised landscape for nature conservation and recreation. Here ecological values take precedence over financial values. Ecoparks is an good example of large forest landscapes with multiple-use management (see color plate 1).

Photo: Rydberg

A third example in this field is the "Sveaskogs Ecoparks" in Sweden. These parks are large continuous forest landscapes prioritizing nature conservation and outdoor recreation and serving as an arena for local nature tour operators, managed by Sveaskog, the state forestry company.

3. *In combination with forest management*:
At a stand level, recreational planning is integrated in local forest management plans in many countries. Recreation and tourism is one important pillar of the concept of multifunctional forestry. Not only are facilities provided but the tree species along forest roads and trails are also selected in order to enhance to aesthetic quality and the experiences of forest visitors. Many examples have been submitted especially from central European countries and from Lithuania. A good example is the Varena Forest Enterprise plan (Lithuania). Here planning, design and establishment of recreational objectives are all included in the forest management plan. The planning is based on sustainable forest development principles and considers social goals and recreational demands in the planning process.

4. *Design of forests with recreation as a major objective*:
Another group of planning processes are those who are – mostly on more detailed level – designed specifically for recreational uses while other forest functions (especially timber production) are not considered or play a minor role. Victory Wood in Kent in England is an example of a new woodland designed for recreation purposes. A series of new forests and woodlands have been planted to symbolise and celebrate the 200 year anniversary of the Battle of Trafalgar in 1805. Each forest is named after one of the ships in the British fleet and Victory Wood is one of them.

Box 2: Design of a recreational forest

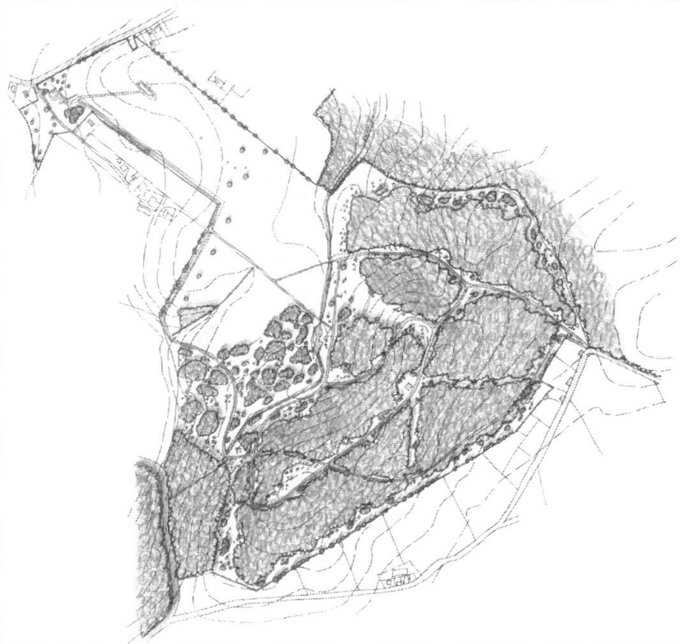

Photo: Bell

Victory Wood (UK) is a new woodland being designed and established primarily for recreation. The landscape design was an important element both in terms of external views to the prominent landscape and also internally, along and from the paths within it. The recreation is aimed at walking, cycling and horses, with some separation of these and, as far as possible, an all-abilities trail leading to the main viewpoints. The landscape is also planned for good nature conservation and to protect archaeological features. The woodlands will be developed over a number of years but access is available straight away.

5. *Planning processes addressed to develop new offers for recreation and nature-based tourism or to solve related conflicts*:
The establishment of new offers in the field of nature based tourism requires separate planning processes starting with a related inventory and evaluation of possible target groups. Here Metsähallitus (the Finnish forest and park service) in Finland is a good example for the development and planning process of a new nature based tourism product called "Wild North".

Box 3: New offers for recreation and nature based tourism

Photo: Erkkonen

Eräsetti Wild North is a business unit run by Metsähallitus, a leading Scandinavian nature tourism operator that sells and provides nature-based tourism and cultural, culinary and wellbeing services tailored to families, businesses and other groups. Nature related activities such as hiking, white river rafting, husky safaris, reindeer programmes as well as quad bike safaris and snowmobile safaris are a part of range of the services. Eräsetti Wild North is Finland's most experienced operator for hunting and fishing trips. Wellbeing from nature and fair trade are the cornerstones in its production.

8.2.2 Steps in the Planning Process

In most cases planning processes can be divided into seven steps. Depending on the different tasks and planning requirements the extent of each step may differ.

1. Inventory of ecological and social conditions
2. Conflict analysis and evaluation
3. Development of vision and goals
4. Development of measures
5. Preparation of the implementation
6. Supervision of the implementation phase (is not always included)
7. Monitoring (is not always included)

These seven steps are illustrated and discussed in the following section.

1. *Inventory of ecological and social conditions*
(Inventories of the fauna or selected species, as well as an inventory of different user groups have to be specified and analysed separately).

The Inventory is a crucial phase for all planning processes. It is the basis of the following evaluation and conflict analysis. Deficiencies in that phase may influence the whole following planning process.

The Europe-wide overview showed that in many countries (see Bell et al. 2009, Sievänen et al. 2008) basic data on recreation in forests is insufficient. Therefore those countries where forest recreation surveys are available have a great advantage and represent the best practice in Europe.

Forest recreational planning and management can benefit greatly from in-depth knowledge about recreation activities, motives, preferences and experiences of conflicts in forest settings. National household surveys provide an overview of the forest recreation of the whole population including also those who do not visit forests often (they might lack opportunities or might have bad experiences). Sievänen et al. (2008) reported that ten countries have made at least one national survey during the past 35 years. They are (in alphabetical order): Denmark, Finland, France, Germany, Hungary, Ireland, Norway, Switzerland, The Netherlands and the UK. The U.K. is a good example of national monitoring of forest recreation. Since the early 1990s, two types of survey have been applied and repeated on a periodic basis: the "GB Day Visit Survey" (GBDVS) and the "Public Opinion on Forestry" (POF).

Box 4: National household survey on forestry

The "Public Opinion on Forestry" (POF) is a biennial survey of public attitudes to forestry and forestry-related issues in the UK. The survey has been conducted since 1995.

In the survey there are sections on forestry management, climate change effects on forests, changes in woodland and a large section on recreational behaviour in forests. There are questions on the frequency of visits, reasons for not visiting woodlands, sources of information as well as attitudes towards the role and use of woodlands (see color plate 2).

Photo: Forestry Commission (2007)

Surveys in local forests reflect use patterns and characteristics of user groups in these areas and provide valuable input to planning and management on the specific aspects of user densities, spatial behaviour and demands in this area. A recent publication on visitor monitoring helps managers to use the most appropriate methods for forest recreation surveys.

Establishing more systematic survey procedures and maybe even comparable data collections across different countries may provide new opportunities for predicting future forest recreation and monitor changes in user patterns, preferences, experiences of crowding and impacts etc.

Box 5: Manual on visitor monitoring

A manual on "Visitor monitoring in nature areas" based on experiences from the Nordic and Baltic countries provides managers with methods, tools and ideas for learning more about outdoor recreation visitors and their use of forests and nature areas. The focus is on how to choose suitable methods, carry out visitor surveys and use the results in management.

Kajala et al. (2007): Visitor monitoring in nature areas. A manual based on experiences from the Nordic and Baltic countries, Swedish Environmental Protection Agency.

Photo: Kajala et al. (2007)

For recreational planning and the consideration of conservation issues in forests and woodlands similar data are often required. Figure 8.2 shows typical information requirements for both aspects.

If there are no data available from national, regional or local surveys alternative sources and a combination of existing data could be used. Figure 8.3 illustrates possible methods and sources for including data on recreation and tourism in the planning process.

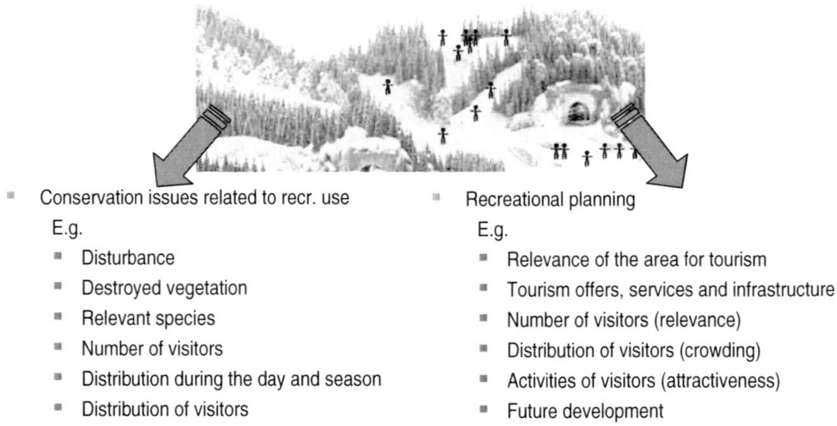

Fig. 8.2 Typical data requirements for recreational planning and conservation issues (Pröbstl et al. 2006)

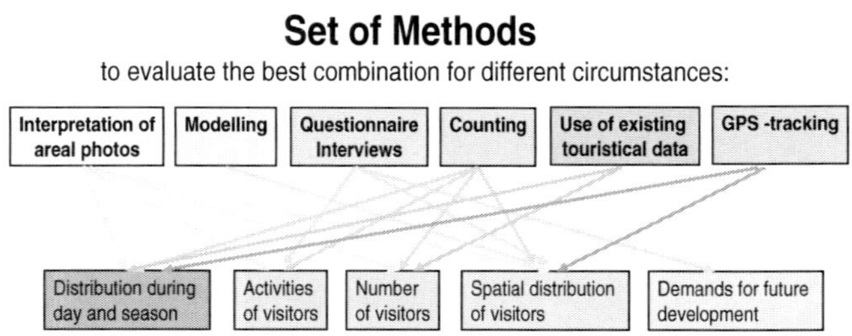

Fig. 8.3 Possible methods and sources to gain and combine data on recreational use (Pröbstl et al. 2006)

2. *Conflict analysis and evaluation*
Any evaluation is based on a specific set of criteria, indicators and possible required standards. The selection of adequate criteria, indicators and standards is closely related to:

- the planning task,
- recreational demands
- existing conflicts with conservation issues
- possible cooperation with different land use partners.

3. *Development of vision and goals*
The evaluation shows possible conflicts or deficiencies. In this phase of the planning process it is necessary to decide on the overall aims. In the case of a specific conflict between different user groups this might be an easy task. Often this phase requires discussion and a need to make trade-offs between conflicting goals e.g.

conservation, recreation and timber production. In this phase of forest planning the involvement of the public, stakeholders and/or communities is in many cases very valuable or necessary. Different methods and good practice for the public involvement are described in Section 8.2.3 in this chapter.

4. *Development of measures*
The decision of main goals and overarching guidelines for the planning process leads in the next step to a development of measures. Also in this step a bottom-up approach is often required e.g. in order to discuss the feasibility and acceptance of proposed or possible measures.

Furthermore, collaboration with the public or selected stakeholders may be necessary in order to prepare for the implementation phase. Another option is to apply new forms of involvement in a public-private partnership concerning the establishment and maintenance of facilities or other contributions.

5. *and* 6. *Preparation of the implementation and supervision*
These phases are necessary if facilities are built and if external companies carry out the evaluation of different offers and the construction of facilities. Phase 5 may also include the calculation of costs, the definition of responsibilities concerning different partners, the selection of steering committee members and the definition of the required framing conditions.

During the establishment of facilities qualified supervision is necessary. The extent and intensity of the supervision depend on the planned project.

7. *Monitoring*
The main task of monitoring is to assess whether the main aims of the planning process have been reached. Therefore monitoring methods should be adapted to the planning aims or should follow the methods which have been used during the inventory.

8.2.3 Public Involvement in the Planning Process

In many cases, especially in urban areas, the traditional top–down approach to forest planning with a focus on timber production is gradually being changed towards a more bottom–up approach of community involvement and a broader focus on diverse forest functions – especially outdoor recreation and nature tourism.

The following examples show different ways of addressing different user groups with different expectations, demands and a varying interest in involvement. One good example of a forest projects with a high level of community involvement is Curringherveld in the Netherlands. The project involves local people in both planning and management of the green surroundings of the community. Terryland Forest Park in Ireland is another example of a high level of direct public participation in both development and implementation. In Iceland, Kjamaskogur is a good example of an Icelandic peri-urban forest recreational area where there has been a close collaboration with the municipality and the local forestry society.

Some projects even take public involvement into the implementation phase in a type of private/public partnerships (PPP). An example is the "Open Forest" project in Iceland where corporate involvement provides funding for making forest areas accessible to the general public and for building recreational facilities.

Box 6: Public involvement in the management of a nature area

Photo: Werkgroep C&HEK, Plaatselijk Belang Kornhorn

The management of the nature area "Curringherveld" in the Netherlands involves many different local groups. The inhabitants of the rural village of Kornhorn were invited to design and manage the Curringherveld area according to their own ideas. This resulted in an attractive area with hiking facilities, school gardens, open-air theatre, etc.

The planning process consisted of "dialogue rounds" with several groups of inhabitants. The implementation of design and management is partly financed by Staatsbosbeheer, partly by sponsorship and partly through the activities of volunteers.

Part of the area has also been "adopted" by the local primary school. Results of the public involvement process are that local people are highly involved and they appropriate the Curringherveld as an essential part of their daily living environment (see color plate 3).

To enhance collaborative processes maps, brochures and guidelines have been developed in several European countries. A toolbox for public involvement in forest and woodland planning has been developed in the UK. It is a collection of different methods of involving the public in planning. A booklet from Flanders/Belgium "Interacting with green space" is a guide which addresses public participation with professionals in the planning and management of parks and woodlands.

Some countries have established dialogues and user councils in relation to forest planning and management and include stakeholders in strategic forest planning.

Box 7: Active participation in the development and implementation of a recreational area

Terryland Forest Park (Ireland) has a high level of direct public participation (through an active steering group) in development and implementation, complemented by a programme of initiatives aimed at directly involving the general public. On the project's steering group there is a considerable community and local NGO representation, a series of high profile public events encouraging direct community involvement aimed primarily at families and young people (e.g. planting days, tree-related art projects), as well as regular articles and newsletters circulated throughout Galway City (see color plate 4).

Photo: Bosbeer

An example is the national "Forest dialogue programme" in Austria. It includes an established process of dialogue as part of the national forest programme using different types of participation including the Internet. Another example is the user councils in all state forests in Denmark which have been working since 1995. These user councils ensure that the population have a greater influence on the management and use of forests and include representatives of various outdoor recreation interests as well as public representatives.

Box 8: Close collaboration between a municipality and a local forestry society

Kjarnaskogur is a peri-urban forest recreational area in Iceland, in the periphery of Akureyri town with 15,000 inhabitants, where there has been close collaboration between the municipality and the local forestry society. This forest was initially planned for recreation from its early establishment on barren, exposed land. The collaboration between the forestry society and the municipality has been the key to the success of the Kjarnaskogur area. The facilities in the area include a network of footpaths, playgrounds, picnic areas and cross country skiing during winter time. The area borders the town, thus being easily accessible to users (see color plate 5).

Photo: Gunnarsson

Box 9: Public private partnerships: "open forest" project (Opinn skógur), Iceland

This picture shows a stakeholder meeting in Iceland – deciding upon what facilities should be established within the forest. In the "Open forest" project the forestry societies collaborate with private corporations to open the forest and build recreational facilities for the public. The project is initiated by the Icelandic Forestry Association, being a country wide network of 60 local forestry societies (see color plate 6).

Photo: Petursson

Box 10: Toolbox on public involvement: involving people in forestry

This toolbox on public involvement in forest and woodland planning has been developed by the Forestry Commission in the UK. This guideline provides general principles and aims of public involvement and gives a broad list of suitable methods. A tool sheet for each method explains which resources and requirements are needed, where to find further information, what is the level of involvement, as well as strengths and weaknesses.

Forestry Commission (2004): Involving People in Forestry. A toolbox for public Involvement in Forest and woodland planning. Forestry Commission, Edinburgh.

Photo: Forestry Commission (2004)

Box 11: Handbook on participation: interacting with green space

This handbook from Flanders, Belgium, with input from Ireland, explains principles and key questions on public involvement in parks and woodlands. It describes a broad range of methods and techniques suitable especially for recreational areas. Application range, costs, advantages and disadvantages are described for each method. The booklet also lists good practice examples in different areas.

Van Herzele et al. (2005): Interacting with greenspace: Public participating with professionals in the planning and management of parks and woodlands, Flemish Government, Nature and Forest Agency, Brussel.

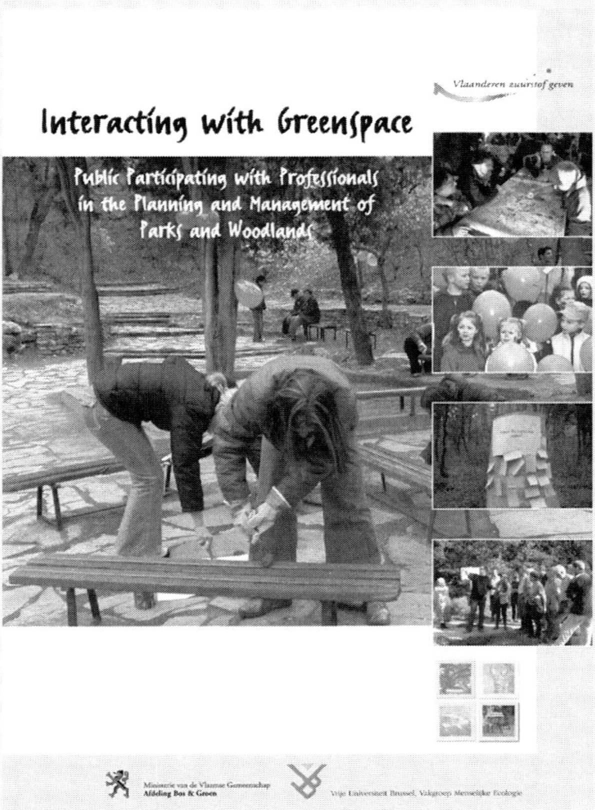

Photo: Van Herzele et al. (2005)

8.2.4 Upgrading Foresters' Knowledge of Recreation Planning

As the role of forests is broadening from timber production to a number of other functions including outdoor recreation, forest planners and managers need to update their knowledge of how to plan for and manage these functions. This need may vary depending on the importance of recreation in a given region, on the ownership of the forest, on the different pressures placed on the forest and the demand by the public.

The expert-based survey described in this book has shown that knowledge and practice regarding forest recreation vary enormously amongst forest managers around Europe (see Chapter 7). Also, the role recreation plays in foresters' education is very different across Europe. The most developed region in this respect is the Atlantic region (especially the UK, Netherlands, Denmark and the Flanders Region in Belgium) since here are the highest pressures on the forest for access and recreational use.

Education in forest recreation needs to cover several aspects and these may differ between management functions and organisational levels. For example senior managers may be involved in recreation planning, resource allocation and forest management at a larger scale such as a forest district or other large unit of management. At this level updated information on planning methods, use of demand and supply surveys, solving conflicts between forest functions, monitoring, economic evaluations of recreational benefits and information on how to manage the forest for increased recreational value is most likely to be required. More junior staff who works directly on the ground are more likely to need information for planning and designing sites and facilities, to know how to manage visitors, to communicate with them and to deal with site-scale conflicts amongst users. If more commercial aspects are involved some skills in marketing and dealing with commercial interests will also be necessary. In most cases, many of these skills are still not a part of the general forestry management or silvicultural curriculum at universities or technical schools.

Ideally, university and technical school programmes should include some aspects of recreation planning, management and design so that graduates gain knowledge about the importance of recreation, how it fits into the changing patterns of leisure and the needs of society, how to take recreation into account in forest planning, how to assess demand and supply and how to carry out simple planning tasks such as zoning of activities.

It is also possible that forestry organisations recruit people to work in recreation who are not foresters by training but already have specific skills in the recreation field and who strengthen the management team at forest unit level. This reduces the need for foresters to have a special skill, although they need enough knowledge to engage in discussions with these recreation specialists. The advantage of employing these people is that they look at recreation from a different perspective than foresters – usually a more people-centred than forest-centred view.

Box 12: Forest user councils

Forest user councils have been established in all the Danish state forest districts to give the population greater influence on the management and use of the forest and nature areas owned by the Danish Forest and Nature Agency. The role of the user councils is to function as an advisory group to the state forest district. The council does not have any power to make decisions. The user council has up to 14 members representing various interest groups including the Danish Outdoor Council, the Danish Sports Organisation as well as publicly elected members. Many of the issues discussed involve outdoor recreation. Summaries of all local user council meetings are uploaded to the homepage www.skovognatur.dk (see color plate 7)

Photo: Lund

Keeping up with knowledge, skills and understanding about forest recreation is part of continuing professional development and many professional organisations such as the Institute of Chartered Foresters in the UK demand that their members maintain their skills by attending events, conferences, training programmes, field trips or demonstrations, as well as reading papers and books.

The Forestry Commission has its own training department and offers training courses run by its own staff or by the other organisations. A recent example concerns UK legislation on disability discrimination applied to Scottish forests. Research into UK legislation, forester's levels of awareness about disability and their needs revealed a lack of knowledge. Therefore, training materials were developed and a one-day training seminar which included a site evaluation with a disabled access auditor was prepared, supported by handouts and presentations.

Box 13: Dialogue process related to forests

The "Forest Dialogue" (Austria) is a dialogue process open for all stakeholders interested in forest concerns. The main focus is solving conflicts in the forest between different uses and interest groups. Within the frame of the process a national forest programme has been developed. The "Forest Dialogue" provides an internet platform where all stakeholders interested in forest concerns can join the discussion and dialogue process. Interested private persons also have the possibility to obtain information on the whole process or to join discussion groups in the internet forum.

Photo: www.walddialog.at

In Ireland over the past few years the Tree Council has promoted a short course on Outdoor Recreation Design as part of the Forest Service support to urban forestry. A two-day course has been delivered in different locations, with presentations and site visits attended by a large proportion of the professional population involved in outdoor recreation management. The course was supported by the book "Design for Outdoor Recreation" (Bell 1997, 2007).

Recreation may also be included in other courses. For example, in Denmark, at the Danish Centre for Forest, Landscape and Planning at the University of Copenhagen (Faculty of LIFE Sciences) and the European Urban Forestry Research and Information Centre (EUFORIC) offer a 5 day course in "Communication in Urbanised Forestry and Green space Management" for an international audience, taught by a number of well-known international experts.

The NOVA network (for cooperation between Nordic Forestry, Veterinary and Agricultural Universities/Faculties) offers a master course in urban forestry and urban greening based at the LIFE centre at the University of Copenhagen and the Swedish Agricultural University at Alnarp. This is a 60 European Credit Transfer System (ECTS) course.

Despite the regional differences in problems, approaches and acceptance, the broad collection of good practices on forest recreation and nature based tourism may provide inspiration and re-thinking of traditional approaches in planning and management of forest recreation.

Box 14: Forest recreation and tourism in Europe: a handbook

This handbook on European Forest Recreation and Tourism gives a good overview of many aspects of forest recreation and nature-based tourism in Europe. It is the main output of COST Action E33. There are sections on the economic and social dimensions of recreation, visitor demand monitoring, and planning at different spatial scales.

Bell et al. (2009): European Forest Recreation and Tourism. A handbook. Taylor and Francis, London.

Photo: Bell et al. (2009)

Box 15: Forest recreation in Ireland – a guide for forest owners and managers, Ireland

This handbook offers practical advice on recreational issues to owners and managers keen to provide for recreation within their woodlands and forests, from forests in the rural landscape to those in and around towns and cities. The Guide contains three chapters dealing with planning and managing for forest recreation, recreation facilities, and recreational issues, with appendices setting out a reading list and contact list. Particular emphasis is placed on practical issues relevant in the Irish context, including co-operation with user groups, issues regarding access and public liability, reducing costs, and dealing with vandalism.

Forest Service (2006): Forest Recreation in Ireland – A Guide for Forest Owners and Managers. Forest Service, Department of Agriculture, Fisheries and Food, Johnstown Castle Estate, Co. Wexford, Ireland.

Photo: Forest Service (2006)

8.2.5 Planning Tools and Guidelines

This section provides an insight in planning guidelines, tools and frameworks. In this context we see an overall influence from the European Union. Certain planning principles and methodological steps are enhanced through different guidelines in the field of precautionary environmental planning (such as the Habitat-Directive, the EIA-Directive, the SEA-Directive and the Water Framework Directive)[2] (Jiricka und Pröbstl 2009):

- Precaution via screening,
- Consideration of alternatives,
- Consultation and
- Mitigation/compensation.

For the planning process there are general guidelines and regulations enhancing the planning quality and also partly influencing costs of implementation. Special planning guidelines have been developed in several countries.

Other publications aiming at guidance on recreation planning are targeted at specific countries, such as the booklet "Forest Recreation in Ireland – A guide for forest owners and managers" adapted to the Irish situation and presenting a wide range of recreational issues such as planning, managing of forest recreation, recreational facilities and recreational issues.

8.2.6 Management Concepts

8.2.6.1 Introduction

Managing forests with intensive recreational use requires specialist knowledge and an ongoing survey and analysis of new developments and changes. In this section we give an overview of tools such as certification and auditing schemes aimed at ensuring continuous management with the permanent consideration of recreational aspects. Furthermore we present management tools which are helpful to deal with natural and social risks.

All management actions require a participatory process as described in the previous section (see Section 8.2.3). The same principles must be considered within the development of management actions, especially if they deal with conflicts.

8.2.6.2 Certification and Auditing schemes

Certification and Auditing schemes both follow the same principles. They try to ensure a sustainable and proper management of the forest over time. They both rely on the use of criteria and indicators to measure the success or failure of management actions and their outcome. The aim is to be able to adjust or to adapt management

---

[2]Environmental Impact Assessment: "EIA- Directive": regulation 85/337/EEC "Habitats-Directive": Directive 92/43/EWG "on the conservation of natural habitats and of wild fauna and flora" "Water Framework Directive": Directive 2000/60EC Strategic Environmental Assessment: "SEA- Directive": Directive 2001/42/EC

Box 16: Certification "living forest", Norway

"Living Forest" is a certification system of sustainable forest management. The main idea is that sustainable forestry will take care of economical, ecological and social values. A Living Forest standard set of criteria have been developed for the certification. Nature tourism is not explicitly included as a partner or as an economic alternative to harvesting, but the experiential qualities and outdoor recreation are important aspects in Living Forests (functional - e.g. paths, ski tracks - historio-cultural, legal, ecological, aesthetic etc.).

Photo: http://www.levendeskog.no/sider/tekst.asp?side=345&submeny=tom&niv2=&menuid=246

Box 17: Certification of ecotourism

There are six basic principles in the certification scheme of "Nature's Best" (Sweden) which have to be are fulfilled by every certified tour operator:

1. Respect the limitations of the destination – minimise the negative impacts on local nature and culture.
2. Support the local economy
3. Make all the operators activities environmentally sustainable
4. Contribute actively to nature and culture
5. Promote knowledge and respect and the joy of discovery
6. Quality and safety all the way.

Photo: http://www.naturensbasta.se

if need be. Certification also serves for providing the public with an indicator of special quality as well as a sign for natural integrity and well balanced use.

The certification protocol "Living Forest" in Norway stimulates forest owners to take care of economic, ecological and social aspects in their forests in an integrated way and to fulfil the requirements of a standardised set of 25 criteria. Outdoor recreation plays a more prominent role in this approach than, for example, in the pan-European Criteria and Indicators on Sustainable Forest Management (MCPFE 2002). In certified "Living Forest" areas management activities have to consider several aspects of outdoor recreation such as:

- maintaining the quality of outdoor experiences,
- maintaining the public right to access and to pick berries and mushrooms (considering the legal situation)
- contributing to appropriate solutions for recreational trails and rest areas
- providing permission and opportunities for environment education (considering the legal situation)

Another certification scheme is "Nature's Best". This is a national certification scheme for ecotourism in Sweden. It has been developed in cooperation with different stakeholders and the public. In 2007, 70 nature tour operators were certified. It includes a number of guidelines on nature-oriented or ecotourism and could probably be highly applicable to other countries. Nature's Best was developed by travel associations, land owners, nature conservation associations, non profit organizations, public authorities, tourist companies and institutions. Nature's Best certifies ethical, high quality nature tours in Sweden.

In Austria, Italy and other Alpine countries an auditing scheme based on the EMAS-II-regulation of the European Commission (Regulation EG/761/2001) has been developed not just for recreational areas but also protected areas. The main advantages can be summarized as follows (Salak 2008):

- Cost saving
- Competitive advantage
- Improving organisational structures
- Minimising risks
- Improving motivation of staff
- Strategic development and cutting edge of the company,

8.2.7 Risk Management Concepts

Risk is a difficult subject in nature and in forests. Three factors influence this subject: natural risks, social risks and the risks caused by the activity.

Natural risks: In the forest there are several natural risks such falling branches or trees, rock falls, avalanches in mountainous areas or fire in southern regions. On the one hand visitors expect to experience nature and wildlife while on the other hand forest recreation should be safe.

Social risks: Some regions face problems with vandalism, destruction of infrastructure and forests being dangerous place for visitors. Vandalism and inappropriate forest use by some socially excluded people are a problem especially the Atlantic countries. Damage to infrastructure is often a difficulty for managers in the Eastern countries As well

Risks caused by the activity: Some recreation activities are more risky than others (e.g. rock or tree climbing, horse riding, mountain biking etc.) which are part of the thrill or experience. Some of these risks also occur when different activities take place in the same area or using the same paths.

All thee aspects are furthermore influenced by the individual risk perception. Actual studies also show that the risk perception of different user groups is often diverse depending for example on sex, experience, age and many other aspects (Arnberger and Haider 2005, McCammon and Hägeli 2007). In the following section management actions according to these four aspects are described and illustrated with examples from different European regions.

8.2.7.1 Management Dealing with Natural Risks

Forest planning and management has to ensure that the overall forest setting and facilities are safe for visitors concerning natural risks such as falling branches or trees, rock fall, avalanches in mountainous areas or fire in southern regions. The duties for forest managers differ depending on the legal background in the different European countries and different cultures. In general, forest roads and frequently used paths are to be monitored and made safe against falling branches, even in National Parks or other protected areas. Naturally, frequently used sites within the forest posing any risk to visitors (e.g. steep cliffs, falling trees etc.) have to be secured and the visitors made aware of any risk. Securing the forest environment also includes avalanche risk management and fire management. No management is required for visitors who leave the official trail network.

Avalanche Risk Management

Forest in mountainous regions has an important avalanche prevention function. The trees hold back the snow on steep slopes or break the impact of slides from higher elevations and hinder the development of extreme avalanche events in general. When developing recreation facilities such as ski slopes, roads, trails, huts etc., the avalanche danger needs to be considered in the planning, design and management of these forest areas. This includes an overall knowledge of avalanche risk zones within the forest. A similar approach is required concerning erosion, debris flow and land/rock slides.

For larger areas there are several good examples with differentiated management plans zoning the area or space in time. In Austria so called danger-zoning plans regulate whether the establishment of infrastructure is allowed or not. Areas with frequent avalanches, debris flow or rock fall are therefore marked in a red colour. Destruction or heavy damage of infrastructure is likely in this zone. These areas are not suitable for additional recreational infrastructure (such as huts, playgrounds, visitor centres). Yellow zones mark areas where there is less danger but damage is still possible.

Fire Risk Management

In the southern European countries as well as rock fall and debris flow, fire is one of these large scale risks. The following example shows a range of concepts for fire fighting. Based on fire planning and zoning systems different management actions can be developed.

In dry regions, precautionary measures and fire fighting methods are very important, in order to protect nature, human lives and also recreational infrastructure. This requires both the right tools and regular training. In Cyprus a fire reel hose system is used especially for the protection of picnic and camping sites. In combination with fire traces and precautionary measures such as fire-look out stations and patrolling this method provides an effective fire protection system.

In wetter climatic zones forest fire is still a topic depending on the season. Picnic sites are the focus of management actions. Establishing visitor facilities such as fire rings is a way for foresters to reduce forest fire risk. Examples of fire rings from

Box 18: Risk index for fire management

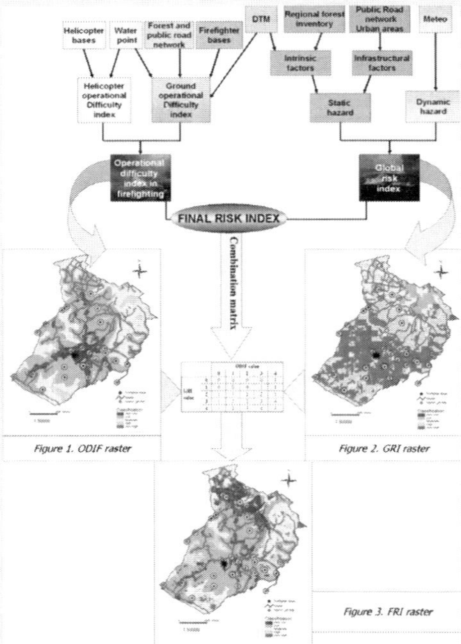

Photo: Regione Toscana – DISTAF University of Florence – CNR IBIMET

The Final Risk Index (FRI) is used to manage and plan the forest fire fighting activities in Tuscany, Italy. FRI is the final result of the combination of different intermediate indexes, the most important are: the Global Risk Index (GRI) and the Operational Difficulty Index in Fire fighting (ODIF), as showed in the figure. The final model is able to provide information at different scales of time and space. The FRI can give information about the daily risk or the seasonal risk. The FRI provides other useful information for a complete planning of the forest fire fighting activity. This model is a very helpful tool for managers in planning the allocation of forest fire infrastructures and features and in the organization of fire fighting. The experience in Tuscany can be considered a good practice for all Mediterranean countries, where the difficulties of logistic organisation of forest fires prevention and fight are often enhanced by a complex mosaic of environments and land forms in relatively small areas (see color plate 8).

Box 19: Fire reel hose system for fire management in recreational forests

Photo: Kazafaniotis

This fire fighting system in the Troodos National Forest Park in Cyprus consists of number of boxes along the perimeter of each picnic site. Each box has a 30 m reel hose (19 mm diam), and can spray water up to 10–15 m, with a maximum pressure of ten bar. Each unit is also equipped with a sluice valve on which 45 mm canvas hose can be fixed, for higher pressure in order to cover longer distances. The system can be also operated by the users of the recreation area in case of emergency (see color plate 9).

Switzerland and Finland illustrate such special facilities to reduce fire risk while allowing people to have fires even in periods of higher risk.

Fire management also includes information, education and awareness campaigns aimed at local residents and visitors aimed at reducing fire risk behaviour. One good example has been developed in Croatia.

Box 20: Fire rings in picnic sites, Switzerland and Finland (see color plate 10)

Photos: Hegetschweiler (above), Paso (below)

Box 21: Promoting fire safe behavior to local residents through community based marketing program "I CARE"

Photo: Marusic

The pilot project "I CARE" or, in Croatian "JA PAZIM" was designed with the aim to change behavior of local residents in relation to burning agricultural waste – the most common cause of forest fires in coastal part of Croatia. The campaign sought to gain residents commitment to the fire safe behavior by following the basics rules of respecting the total fire ban over the summer months as well as behaving responsibly and according to legislation in the rest of the year – especially notifying fire-brigades on the intention to burn in the open by using the emergency phone number. Target groups were local residents involved in framing or agriculture, primary school children who were instructed to discuss the subject at home and the public in general. A community based social marketing approach was adopted. Means of direct communication methods included household visits by members of local fire brigades, presence at community events, education in schools with competition for best essay or drawing on the topic and distribution of message through the local church leaders. The set of "I care" promotional material was distributed including brochures, posters, stickers, t-shirts, picture book and awards. The local radio, television and newspapers were also used to reach wider population and increase receptiveness to the campaign message and more targeted community-based activities.

8.2.7.2 Management Dealing with Social Risks

Vandalism Risk Management

Some regions face problems with vandalism and destruction of infrastructure. Vandalism may be experienced at access points where cars may be broken into, toilet facilities vandalised or litter bins destroyed. This may also be caused by people who do not use the forest for recreation. However, forest visitors may also vandalise trees by carving, breaking branches, uprooting plants, digging, vandalising signage etc. Information on what is allowed and awareness-raising campaigns, fines for violators etc. may be tools to prevent some of the vandalism. Positive encouragement tends to work better than too many prohibition signs.

Box 22: Alternative risk and vandalism management approach

A booklet has been published by the Forestry Commission in England using an alternative risk and vandalism management approach.

It provides help to managers to decide when to act on self-made structures in the forest constructed by recreational users, such as children or teenagers. The booklet provides guidance on rope swings, dens, tree houses and fires, all typical structures built by such groups (see color plate 11).

Photo: Forestry Commission

There is a gradual line when the actions taken by the visitors in the forest can be characterised as "vandalism". Some vandalism may be motivated by an underlying urge to interact with nature by constructing playthings, testing boundaries etc. Allowing some space for this may be a vandalism reducing measure in itself. A good example is an alternative risk and vandalism management approach from the UK.

8.2.7.3 Management Dealing with Risks Caused by Activities

Some recreation activities contain more risk than others (e.g. rock or tree climbing, horse riding, mountain biking, back country skiing, free ride skiing etc.) which are part of the thrill or experience.

Back country skiers, snow shoe walkers and free ride skiers face the danger of avalanches as they do not use prepared ski slopes. Information, awareness-raising campaigns and training for these user groups are a key factor in their avalanche risk management. The alpine clubs and some private providers also offer avalanche risk trainings. The Dachstein area in Austria is a good practice example for additional awareness raising.

Box 23: Rising awareness for the risks of leaving ski slopes

The Dachstein area in Austria is very popular for back country skiing and free riding. Most people use the cable car for a part of the ascent and leave the ski slopes afterwards. When leaving the cable car plateau, a signal indicates to the backcountry skiers whether the security equipment (the avalanche detector) is turned on. This is a good example for very popular free riding areas.

Photo: Wirth

8.3 Detailed Management Actions and Specific Solutions

8.3.1 Adaptive Management and Conflict Resolution

Veronika Wirth[3] *and Berit C. Kaae*

With contributions from: Peter Breman, José Manuel Castro, Marcel Hunziker, Savvas Kazafaniotes, Gudrun von Langehove, George Pattichis, Pieter Roovers.

8.3.1.1 Introduction

Many types of conflicts can arise in relation to recreation in forests. There are several categories, levels and dimensions (see e.g. Jacob and Schreyer 1980, Ammer and Pröbstl 1991, Manning 1999, Schemel and Erbgut 2000, Cole and Monz 2002). Conflicts in relation to recreation can also have a multi-dimensional nature, e.g. conflicts between skiers and snowmobilers might be related to the different values of each group, or to conflicting recreational goals in their desired recreational experience (e.g. quiet and solitude versus speed and excitement), or the skiers' feeling of insecurity along the track. In this COST Action, conflicts have been categorised as follows: conflicts between recreation and nature conservation, between recreation and different land uses (forestry, hunting) or conflicts within recreational use, between different user groups (e.g. hikers and bikers along a shared forest track) and due to crowding pressures.

It is not the aim of this chapter to present an exhaustive compilation of recreation-related conflicts and management approaches in forests but to present a range of good practice examples of how to solve existing conflicts, adapted to the variety of European forest settings. Possible solutions cover a broad range:

Adequate *recreation planning* can prevent conflicts developing (see Section 8.2), but in practice this is not always possible.

Spatial and temporal solutions are often applied for recreational conflicts. Through spatial zoning and separation of various uses visitors can be guided to dedicated areas and so conflicts related to wildlife and natural habitats and other land uses can be reduced or avoided. These solutions are also used to solve conflicts between different user groups, for example by providing different spaces for different uses. Spatial solutions include both restrictive and soft management actions (see Chapter 7); the good practice examples in this section include zoning of areas, closing areas for recreation and offering attractive infrastructure to guide visitors.

Temporary solutions imply that areas are closed for recreational use during a certain period of the year. This can be an effective solution to protect sensitive species during the breeding season, e.g. birds.

[3]University of Natural Resources and Applied Life Sciences, Institute of Landscape Development, Recreation and Conservation Planning, Peter-Jordan-Straße 82, 1190 Vienna, e-mail: veronika.wirth@boku.ac.at

Voluntary arrangements with specific activity groups go one step further. In a participatory and target group-orientated approach management actions are agreed between managers and representatives of affected recreational groups. This chapter includes good practice examples of agreements to avoid disturbance of habitats and species and conflicts between user groups.

Guiding and awareness-raising solutions aim to increase awareness among users and to convey a way of behaviour that reduces impacts to sensitive habitats and species. Good practice examples include informing and guiding recreational users visiting sensitive forest environments.

This chapter presents examples of both approved tools applied in recreation sites all over Europe and of new and innovative solutions that have been developed and tailor made for complex problems.

8.3.1.2 Conflicts Between Recreation and Nature Conservation

Recreational activities have increased in importance especially in forests over the last decades but they may also have an impact on natural habitats and disturb wild species (see e.g. Cole et al. 1997, Manning 1999, Schemel und Erbguth 2000, Ingold 2005). Also in the European comparison analysed for this book the conflicts between recreation and nature conservation are the most important type (see Chapter 7). In this section a range of examples of how to solve this conflict are presented. The majority of the examples relate to spatial and temporal solutions. There are also solutions that include the recreational users in their development, such as voluntary agreements and guiding and awareness-raising solutions.

Spatial Solutions

Spatial solutions aim to create separate areas for the purpose of recreation and nature conservation. Zoning of forest areas is a useful planning and management tool to provide, on the one hand, areas with attractive potential for recreational use capable of coping with sufficient visitor numbers and on the other hand areas where nature conservation aspects have priority. One good example is the adoption of beach plans on the Aquitaine coast in France, where high numbers of visitors are managed in a sensitive environment.

Fencing off areas within forests is an effective management action to protect sensitive areas, especially if they can be severely affected by trampling such as peat and swamp areas and where other methods for diverting pedestrians away from the area have failed. An example is the Troodos National Forest Park in Cyprus where endemic peat grassland has been fenced off in a heavily used recreation area in order to protect this sensitive habitat.

Box 24: Beach plans – zoning and channelling activities

The use of Beach Plans (France) is a successful policy to manage a high number of visitors in a sensitive forest and coastal environment. They are the result of a wide-ranging and close collaboration among all public stakeholders involved in the tourist development and the environmental preservation of the Aquitaine coast. More than 30 Beach Plans have been implemented over the last 25 years. Car parks are laid out behind the mobile dunes and integrated in a shady forest. The public are directed towards supervised beaches by wooden boardwalks, a path ensuring both the comfort of the users and the protection of the coastal dune. A cycle trail network also gives the opportunity to go cycling to the beach and to enjoy the pine forest landscapes.

Photo: ONF

A promising management approach to solve conflicts between recreation and nature conservation is to provide attractive infrastructure for recreational use and for environmental education, so that visitors can be guided towards those areas that are designated for recreational use and away from sensitive areas. The Bertiandos Lagoons Protected Landscape (Portugal) is an example where nature conservation has been successfully combined with the development of recreational opportunities to manage visitor flows. Local involvement has been part of the process.

Box 25: Fencing off sensitive areas

The Troodos National Forest Park in Cyprus hosts a habitat of European interest, the "Peat grasslands of Cyprus" which are protected by the European Habitats' Directive (Directive 92/43/EEC). It is a rare habitat with rare plant species, endemic to Cyprus. The site is mainly threatened by the pressure of recreational activities. The protection of peat grasslands is achieved by erecting fencing for habitat and species conservation. The surrounding area is a Natura 2000 site and the fenced section is a Nature Reserve, very close to a picnic site (canopied area in the background).

The work was funded by a LIFE project and it was implemented by the Forestry Department. The materials used for fencing fit to the environment (see color plate 12).

Photo: Tsintides

Some recreation activities can disturb plants and soil. Different forests habitats and wildlife species within the forest have different degrees of resilience towards these impacts. The degree of impact and the regeneration time following the impact will vary according to a range of ecological factors, such as surface, moisture, soil and vegetation type or length of growing season.

One management action in such sensitive habitats is to provide wooden boardwalks to prevent impacts to vegetation and soil and to channel visitor flows. Neigembos in Flanders/Belgium is one example where a wooden boardwalk has been constructed to prevent impacts to a sensitive part of an intensively used forest

Box 26: Combining protected landscapes with development of recreational infrastructure

Bertiandos lagoons are a wetland that is designated as Natura 2000 site in Portugal. This protected area consists of a riparian system and an agro-forest mosaic of high conservational value. It was classified as a Regional Protected Landscape in 2000, following the proposal of the municipal authority, based upon faunal and floristic data. The Protected Landscape offers an environmental interpretation centre with information about the area, five nature trails and three cultural trails, as well as a campsite, lodge, fishing sites and viewing towers that offer observation hides to allow visitors to observe the wildlife without disturbing their activities.

Photo: Castro

in an urban area. Another example is the Evo Hiking Area in Finland, where a wooden boardwalk leads visitors through a bog forest in a comparatively remote area.

Temporal Access Restrictions

Some recreational activities in forests can disturb wild animals most severely during their breeding season. Birds are especially likely to be disturbed. Temporary access restrictions during this time are very effective at preventing disturbance. One example is from Denmark, where access to areas where eagles nest is restricted during the breeding season.

Box 27: Wooden trails in Belgium

In 2003 the Flemish Government constructed a wooden boardwalk to enhance accessibility to the "Neigembos" area (especially during wet periods) and to protect the sunken road from harmful use and from the creation of undesirable new routes. The wooden construction has a width of 1 m and a length of 220 m, and is constructed on posts. Due to the specific terrain, environmental and user constraints (permanently wet, dark during the growing season, intensive use) a well-designed and sustainable construction was essential. Therefore the material used was Massaranduba (paraju), a FSC labelled durable timber. The path has to deal with a 24 metre difference in height along its length (see color plate 13).

Photo: ANB

Box 28: Wooden trails in Finland

In the recreational forest of the Evo Hiking Area, a wooden boardwalk has been constructed in a boggy forest area. Evo is one of the largest continuous forested areas in southern Finland. In the hiking area recreational use is the main purpose but there are small protected areas within it and small scale timber harvesting is also acceptable when it fits with recreational use.

The trail consists of two planks of local timber laid side by side. The trail is rather narrow but allows two visitors to pass. At certain viewpoints there are larger platforms where visitors can pass more easily. The forest is a rather wet and sensitive area, so that people do not to follow the boardwalk and are very likely walk freely in the forest, thus protecting it (see color plate 14).

Photo: Wirth

Box 29: Temporal access restrictions

Photo: Caspersen

When a White-tailed Eagle (*Haliaeetus albicilla*) in 2009 started nesting in Nejede Vesterskov forest in Denmark, temporary access restrictions were established. To protect the birds from visitor disturbances in an intensively used recreational area, a section of the forest has been closed off and guidelines on non-intrusive behaviour in the rest of the forest are posted by the Forest and Nature Agency in North Zealand in cooperation with the local Bird Watching Society.

The White-tailed Eagle is a rare breeding species in Denmark. Approximately 21 pairs of nesting bird couples were identified in 2008 and they collectively raised 29 chicks to fly from the nest. The bird is territorial and keeps the territory for life so temporary access restrictions in the breeding season can help. The hatching of a chick has now been reported.

Voluntary Agreements

Encompassing these temporal and spatial solutions described above are solutions that include the recreational user groups in their development and implementation. Voluntary agreements with specific user groups imply solutions for a given area are developed as a consensus between the representatives of the activity and the organisation responsible for the area (e.g. nature conservation administration, administration of a protected area etc.). As part of the deal, the users commit themselves to implementing the agreement. An example of such an agreement is the integrated climbing management scheme applied in the low mountain ranges in Germany, areas that are especially important for birds and therefore often dedicated as protected areas.

Box 30: Voluntary agreements – climbers

In various climbing areas in Germany there is an agreement between the Alpine Club and nature conservation administration setting out different zones:

- closed for climbing
- priority nature conservation (restrictions are possible e.g. temporarily closed, restrictions on new routes and bolts. . .) and
- priority climbing (climbing is possible without restrictions. Establishing new climbing routes is possible).

Zoning on access agreements and regulations is marked in a consistent way. Access agreements are communicated through a database of the Alpine Club and on site. In most climbing areas there is also a responsible person who acts as the primary contact person. These visitor management actions are voluntary but if the agreement is not kept it would also be possible to implement regulations on a legal basis (see color plate 15).

Photo: Taczanowska

An advantage of these solutions is that agreements are usually more easily accepted than restrictions or even completely closing areas for recreational use. They also include the views of the concerned user groups and thus ensure the best possibilities for retaining the remaining recreational opportunities (e.g. very traditional or favourite routes remain accessible for recreation). Thus a compromise between conservation and recreational goals can be achieved. It is usually advisable

Box 31: Voluntary agreements – water related sports

At three Bavarian lakes (Starnberger See, Ammersee, Chiemsee) a written agreement has been reached between the user associations and the nature conservation administration on how to manage water sports (sailing, surfing, paddling, canoeing) including the following topics:

- Zoning of the area to protect shallow water areas as habitat for birds
- Temporal restriction for water sports (closed from 1st of November until the 31st of March)
- User associations and their members ensure that the agreements are kept and support information campaigns about this agreement. Unorganized users are informed about the regulations by leaflets and on site. (www.natursport.info)

Photo: Wirth

that the agreement is communicated in an appropriate way. A good example is the selection of persons that are responsible that the agreement is adhered to in a given area (e.g. persons that are responsible for a certain climbing area). It is also important to include monitoring aspects and to have sanctions which can be inposed if the agreement is broken (e.g. that the area will be closed to the offending recreational group). This ensures that all users of the area are motivated to keep to the agreement as not to lose the recreational opportunities they enjoy. A comprehensive compilation of further agreements in German and French can be found at www.natursport.info.

Guiding and Awareness-Raising Solutions

Guiding and awareness-raising solutions aim at informing visitors of habitat requirements of wild species and the possible impact of their recreational activities if they are carried out in a careless manner. Often visitors are not aware of these impacts. The aim of these guiding and awareness-raising solutions is to show visitors how to reduce their impact on wildlife.

This type of management is especially suitable in the case of new and increasingly popular leisure activities – such as snowshoeing – that are often carried out in sensitive wildlife habitats. The snowshoe project in Ibergeregg (Switzerland) is a good example of how to reduce disturbance by active guiding through information and infrastructure. Snowshoe walkers particularly like semi-open forest, which is at the same time the habitat of rare species. The challenge in managing such an activity is to protect sensitive and quite often fragile habitats for wildlife and habitat conservation and diversity.

Other examples are guided snow shoeing tours in the Chartreuse mountains in France, where messages about the natural environment are part of the overall experience and the rangers in Gesäuse National Park in Austria, who raise awareness amongst backcountry skiers about a sensitive capercaillie population and who take care that the newly established trail regulations are kept.

To be successful in solving these conflicts between recreation and nature conservation it highy advisable that the marking of trails should be combined with active communication to convince recreational users that they should cooperate. Communication should not be limited to on-site information. The planning of a mountain trip often starts earlier, and so should the communication – via websites, for example.

8.3.1.3 Conflicts Between Recreation and Land Uses (Including Hunting)

Recreation is one of several forest functions. Recreational conflicts may for example be related to timber production where even-aged forest stands are favoured from a timber production perspective while recreationists may prefer more uneven-aged forests with diverse tree species. Timber harvesting frequently causes conflicts as aesthetic and experiential recreation values may be strongly reduced. To avoid this it is advisable to take into account scenic beauty issues (see next section). This conflict type is very important especially in the Nordic countries (see Chapter 7). A good example of how to tackle problems related to timber harvesting in an intensely used recreation area is the urban forest of the metropolitan area of Helsinki.

In addition, conflicts between recreational use and hunting may occur. This is especially likely in densely populated areas where recreationists affect wildlife. There may also be the risk of accidents due to hunting.

A good example from Austria, a country with a very long hunting tradition and a strong hunting lobby, is the integrated management in the Wienerwald biosphere reserve. In this project several land user groups have worked together to find solutions in an area that is very important for both recreation and hunting.

Box 32: Raising awareness for wildlife disturbance by special user groups

Photo: Freuler

The snowshoe project in the Ibergeregg (Switzerland) aimed at reducing recreational conflicts between snowshoeing and wildlife by defining a limited number of trails in order to channel visitors away from sensitive areas. The trails were defined in cooperation with all concerned stakeholders (foresters, tourism managers, authorities, conservationists, game wardens, etc.). Four trails of different length and difficulty were marked. At the start points, visitors were provided with trip instructions and ecological information on protected areas, sensitivity of wildlife during the winter period and the need to stay on the trail to reduce disturbance of wildlife. The impact of these actions was evaluated scientifically. The results showed that steering snowshoe walkers towards specific areas is successful if visitors are provided with additional ecological information. The project succeeded in reducing the impact on the protected area significantly (see color plate 16).

Box 33: Guided tours for special user groups

In the Chartreuse mountains (France) guided snowshoeing tours at night in winter are organised by foresters of the French Forestry Office (ONF). While the visitors are looking for animal footprints and listening to the forest sounds in the night, messages about the forest and its wildlife can be given to the group.

The tour ends in a local mountain restaurant and the snowshoes are rented at a local place which strengthens the local economy.

Photo: ONF

8.3.1.4 Conflicts Within Recreational Uses (Crowding, User Conflicts)

Social conflicts may occur when recreational activities and experiences are disturbed by different recreational groups. The conflict is often one-sided (asymmetrical) meaning that only one of the groups is disturbed (e.g. hikers may feel disturbed by mountain bikers while the mountain bikers do not feel disturbed by the hikers or non-motorised users are disturbed my motorised users). It is often linked to conflicting recreational goals – those seeking peace and quiet may feel disturbed by those seeking thrills and excitement or by those seeking the forest to socialise within the group (cp. e.g. Manning 2007, Watson 1995, Mann 2006, Arnberger und Haider 2005, Vittersø et al. 2004, Wöhrstein 1998).

Conflicts may also occur within recreational groups, e.g. due to crowding. This is particularly the case in forests close to urban areas or in forests with very unique features which can attract a large number of people. Crowding is linked to the number

Box 34: Ranger system to raise awareness and ensure existing rules are kept

The Gesäuse National Park in Austria hosts an endangered habitat of capercaillie (*Tetrao urogallus*). At the same time the area is very popular for backcountry skiing. To avoid disturbances of the capercaillie population a traditional ski route has been modified: the part that passes through the capercaillie habitat has been closed and a deviation is signposted. Unfortunately the old path is still very popular with the skiers because it is shorter and more attractive.

Thus, on popular days, rangers inform skiers at the start of the closed branch about the sensitivity of the grouse population and ask them to use the new route (see color plate 17).

Photo: Thaller

of other people in a forest setting, but also to the expectations of the visitors. A feeling of crowding occurs when the number of other visitors is much greater than expected and when the higher number negatively affects the recreation experience and the underlying goals (Shelby and Heberlein 1986, Manning 2007, Arnberger and Haider 2007, Vaske and Shelby 2008). A birdwatcher may feel crowded with only a few other visitors in the forest, as they interfere with their goal of viewing birds undisturbed. A Nordic walker may not feel crowded even when many others are close by, as the activity – for many – has a social objective.

In the European comparison these types of conflict tend to be most important in densely populated countries and areas. Management initiatives to reduce crowding can include providing an appropriate capacity of facilities in relation to the general number of visitors (parking spaces, trails, toilets, huts etc.). Even economic means

Box 35: Timber harvesting in intensely used recreation areas

Helsinki Central park is the urban forest of Helsinki with 700 ha forest and 100 km of trails; the total area is 1,000 ha.

The forest is managed as continuously forested recreational area. Recreation and biodiversity protection are the main functions. However, timber harvesting also takes place in this area from time to time.

When harvesting is planned there is a broad information campaign (announcements, internet, signs) and a telephone hotline which recreationists can call for specific information. The information campaign starts already when the management plan is done and continues during the whole implementation process until the harvesting is finished.

Photo: Saukkonen

such as fees can sometimes be used. This is a common management tool in North America but in European forests this is rather an exception as in most countries there is no such tradition.

Management actions to reduce conflicts between visitor groups usually aim at spatial solutions, such as zoning of noise activities away from quiet ones, and develop these solutions in a collaborative manner or through voluntary solutions such as codes of conduct.

Box 36: Integrated sustainable wildlife management

Photo: Heckl

Wienerwald Biosphere Reserve is a peri-urban forest with overlapping land uses. The area is very important for recreation for the inhabitants of the city of Vienna and is at the same time a very traditional hunting ground with many small sized hunting districts.

This project developed concepts, indicators and tools for sustainable wildlife management with a participatory study approach. To analyse overlaps and interfaces of various land uses (hunting, forestry, farming and recreation) a number of principles and criteria were developed. A self-administered online-tool was established enabling hunters to assess the sustainability of their hunting. Out of the analysis recommendations for management and monitoring are be developed. Regarding recreation and hunting, the project underlines the importance of exchanging information between the stakeholder groups and coordinating visitor and hunting management.

Another outcome is raising awareness and building a kind of code of conduct among the representatives of the land users so as to encourage them to behave in a sustainable way and to avoid conflicts with other uses (see color plate 18). The criteria for sustainable hunting are available in English: http://microsites.umweltbundesamt.at/fileadmin/inhalte/chm/pdf-files/Englische_pdfs/Sustainable_*Hunting._Principles__Criteria_and_ Indicators._Revised_and_extended_edition_2006.pdf*

Box 37: Separate trails for different user groups

Photos: Kaae (above), Danish Forest and Nature Agency (below)

In Jagersborg Dyrehaven (Denmark) horseback riders have a separate trail system with an adequate surface cover for horses. Thus riders have better recreational experience as quicker riding is possible and there are less disturbances from other users. Other recreational users share wide gravel or paved roads without being disturbed by riders.

To mark different trails for different user groups the Danish Forest and Nature Agency has developed a standardised signage system for all nature areas including forests.

The signage system is based on a wooden post combined with pictograms, signs of different size, route signs, etc. On state land the posts are red, on municipal land often brown and on private land black (see color plate 19).

Spatial Solutions

Trails provide a very basic infrastructure in most forests. Providing separate trail systems for different user groups helps to avoid user conflicts and thus better recreation experiences. Examples are the creation of different tracks for horse riders as in Jagersborg Dyrehaven (Denmark) or the separation of trails for mountain bikers and hikers in Uetliberg (Switzerland).

Box 38: Dedication of forests to special uses – dog forests

Photo: Schiøtz

About 137 forests (or parts of the forest) in Denmark are designated "dog forests" where dogs can run freely under the supervision of the owner. These are often located near urban areas. Dog forests are fenced and equipped with self-closing gates. Signs inform the dog owners about the rules of use. Benches and tables, litter bins and "doggy bags" are provided. Sometimes a part of the area is designed for dog training in cooperation with a local club. The provision of car parking facilitates dog owners from a broad hinterland to reach these forests. An internet based map as well as pamphlets provide information on the dog forests in Denmark (see color plate 20).

Another frequent conflict between user groups is between dog owners and other recreational groups. Dogs like to run about without a leash but other users are likely to feel disturbed by the dogs. In heavily used recreational areas this may severely disturb the recreational experience of many users, especially if there is a lot of dog mess left on the ground. One good example of solving this conflict is the "Dog Forests" of Denmark.

A good practice example of how to handle conflicts between user groups is from Uetliberg in Switzerland. A survey among Swiss foresters showed that conflicts occur between hikers and bikers on single trails in alpine areas and even more often in urban forests, where the number of visitors is very high. A survey among forest

Box 39: Integrated management: hiking – biking conflicts

The project "Hiking and Biking at the Uetliberg" (Switzerland) focuses on resolving conflicts between different user groups in an urban forest with high visitor numbers. Several approaches were combined and well communicated to visitors. The interventions aim to reduce conflicts between hikers and cyclists. All user and interest groups were involved in the planning process.

The ban on cycling on forest roads was removed to increase acceptance of "normal" cycling. Illegal downhill trails were closed to reduce disturbance and erosion (82% of all visitors agreed with this); one special bike trail was built, to separate hikers and off-road bikers (86% agreed). Bike transportation with the train was stopped to prevent frequent downhill runs at speed and also heavy bikes. The whole project was communicated at a media conference to inform forest visitors. Signposts inform cyclists about the trail and its rules (see color plate 21).

Photo: Freuler

visitors at the Uetliberg confirmed this at a local level. Seventy percent of the forest visitors felt at least sometimes disturbed and endangered by cyclists (mainly because of fast riders). Some cyclists used a train leading to the mountain top, for frequent and fast downhill runs on trails, single trails and roads. To resolve this conflict, the city of Zurich implemented a combination of several methods: to reduce user conflicts between bikers and hikers separate trail systems were established. General rules were also agreed. In the whole process all affected user group were involved.

To be successful in solving these user related conflicts, it is important that all concerned user groups are involved into the planning process. Even small groups can develop high resistance, if they feel ignored. To ensure acceptance, any restrictions should be communicated and justified to the affected forest visitors.

Voluntary Agreements

Some new recreation activities involve relatively large groups of people that may cause disturbance to other users. To avoid conflicts between user groups it is advisable to make arrangements with these groups.

Such arrangements should account for the amount of people conducting this activity as well as both the temporal and spatial zoning. In some countries these may require individual arrangements and permission from the forest owners. The appointments can be made formally or informally, and it is advisable that the arrangements are monitored. Examples include arrangements with Nordic Walkers in the Netherlands and a code of conduct of mountain bikers in Exmoor National Park.

Box 40: Arrangements with specific activity groups

In the Netherlands, a covenant of various institutions (a binding agreement about the performance of a shared action) has been used to establish a common strategy to accommodate Nordic walking as an activity. This includes design and management of Nordic Walking paths, special activities and commercial user fees, rules of conduct, publicity and promotion (see color plate 22).

Photo: Kaae

Box 41: Code of conduct of specific activity group

Exmoor National Park is very popular for off-road cycling and provides a good network of tracks. To reduce conflicts between user groups a Code of Conduct for Cycling was developed:

- Ride only on permitted tracks
- Give way to walkers and horses; control your speed and avoid bunching when travelling in large groups.
- Announce your presence when approaching horses and walkers from the rear, to avoid startling them, ideally with a bell, as it is more friendly and less startling than a shout!
- Minimise your impact on the environment by avoiding excessively worn tracks especially during wet weather.
- Skidding during breaking is very destructive, particularly on the delicate moorland soils, so brake sensibly.
- Follow the Country Code and use your common sense. Be responsible

http://www.exmoor-nationalpark.gov.uk/index/visiting/activities/cycling/code_of_conduct.htm

Photo: Wirth

8.3.2 Supply for Recreation in Forests

Simon Bell,[4] Art McCormack and Berit C. Kaae

With contributions from: Olgirda Belova, Jan Blok, José Manuel Castro, Dora Drexler, Birgit Elands, Joel Erkkonen, Tessa Hegetschweiler, Savvas Kazafaniotes, Barbara Mariotti, George Pattichis, Veronika Wirth, Roger Worthington

8.3.2.1 Introduction

Forest recreation, by its nature is about visiting and experiencing forest landscapes. Thus, while the infrastructure such as parking, picnic sites and trails is needed to enable visitors to experience the forest, this should always be secondary to the landscape. The forest itself may vary in character and already provide a diverse and attractive experience. The different European regions demonstrate different characteristics. For example, the extensive natural forests in Lapland provide quite a wild and remote experience where the visitor can escape from other people and where the character of the forest may be wild and feel untamed. In central European countries such as Poland the forest at Białowieza also has a primeval feel, dominated by large trees and a considerable space beneath the canopy, while other forests may be more intensively managed with many signs of human activity such as fellings and geometric rides. In north-western Europe many of the forests are of plantation origin and may comprise densely planted stands of non-native conifers which are planted in straight lines. These forests need considerable work using various forms of forest management if they are to become attractive for recreation. Where new forests are being planted there is scope to plan and design them so that they contain a lot of diversity of open spaces and species, for example, from the outset.

Since the forest is the most attractive aspect of the purpose of the visit for recreation all infrastructure should be planned and designed so as to be subservient to the forest and to facilitate the enjoyment of the visit. Thus the use of materials, the form and layout of facilities and other objects should be carried out with this central philosophy in mind. Another aspect to be considered is the need to provide a contrast between the forest landscape and urban landscape from where most visitors are likely to originate. This means avoiding urban forms, styles and materials in favour of more natural ones so as to help visitors use the recreation experience to wind down and de-stress themselves.

8.3.2.2 Design of Forests

Traditionally the primary focus for forests has been on timber production, resulting in the predominance of one or a few commercially productive tree species and

[4]Edinburgh College of Art, OPENspace Research Centre, Lauriston Place, Edinburgh EH3 9DF, UK, e-mail: s.bell@eca.ac.uk

creating a monotonous forest from a recreational point of view. The multitude of functions that forest provide, however, are increasingly recognised. For example, the ecological and recreational functions of forests are coming more into focus and, among other forest functions, are being appreciated for the experiential and aesthetic aspects of visitor enjoyment.

Box 42: Forest design

An example of a book which aims to ensure that the forest landscape is protected or enhanced during all kinds of activities. Forest landscape aesthetics need to be considered at the same level as other aspects of the forest environment, and design principles should be applied as part of forest planning.

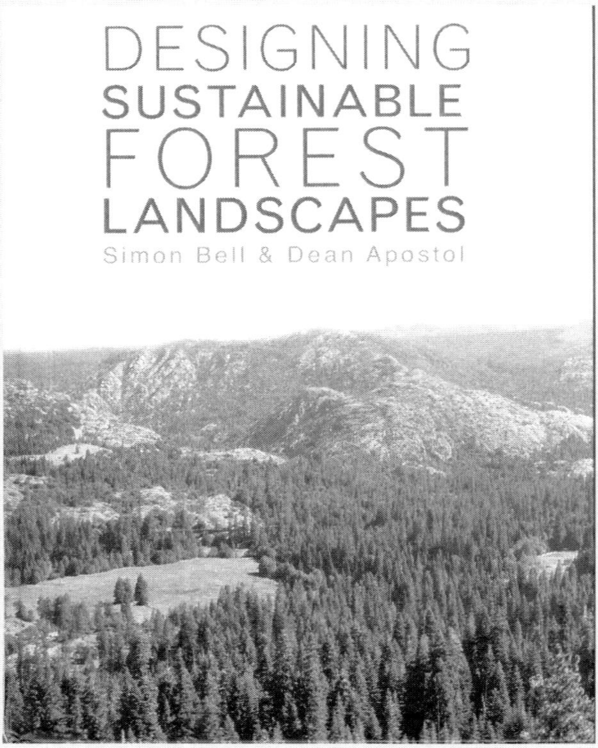

Photo: Bell

Box 43: Recreational forest

Some new forests can be designed with recreational use in mind form the outset, so that walking routes, viewpoints, recreational spaces and opportunities for interpretation are built into the design (see color plate 23).

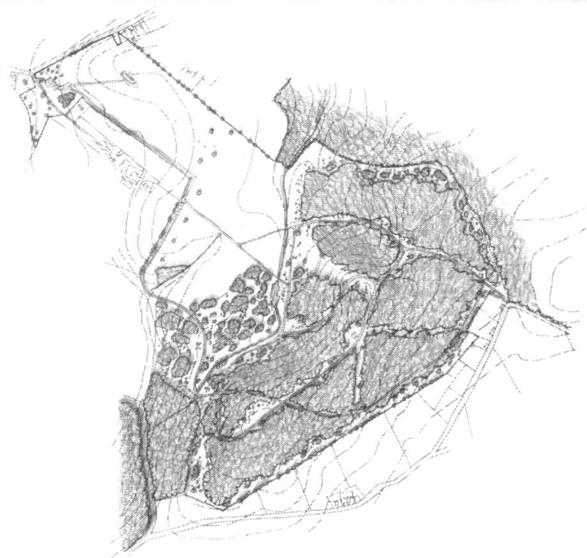

Photo: Bell

The aesthetic of the forest may be experienced in two ways: as part of the wider scenery of the area within which recreation or tourism is taking place. This is especially important in countries with many hills and mountains offering wide panoramas of the landscape. The forest may be a dominant element or a smaller part of a wider and diverse scene. The layout of the forest, of elements such as felling coupes, forest roads, fire breaks and other managed features may create unattractive aspects which good design can refine. This is especially important in places where the forest is intensively managed and of plantation origin. Forest design principles have been developed over the last 40 years and widely applied in places such as the UK and Ireland where such problems have been most acute.

In less hilly or mountainous landscapes the forest is primarily experienced from within, from roads passing through it, from views across lakes and from trails winding their way among the trees. Aesthetic enhancement can be achieved through the manipulation of the forest in terms of increasing the diversity of the forest structure, species composition and use of open space. In practice, this is realised by creative

Box 44: Recreational forest

In forests where there is a strong recreation function, visitor experience must be given particular consideration. An example of such a forest is the Avondale Forest Park, County Wicklow, Ireland, which combines a diverse range of species and spatial (forest structure) experiences at a scenic location by the Avoca River (see color plate 24).

Photo: McCormack

use of silviculture, species selection, stand structures and management actions as thinning, pruning and felling. The degree of diversity seen by recreational visitors will depend in part on how they experience the forest. For example, when driving through it in a car the landscape appears to pass by quite quickly, so that the diversity has to be at a larger enough scale that it can be seen. People travelling by bicycle or on foot can perceive more detail along path edges, or can see a view glimpsed through a narrow opening which would be missed at speed. Thus there are many levels of diversity to be included, ranging from the landscape as a whole viewed from outside to the smallest detail observed by a visitor sitting on a bench. The forest area used by visitors should therefore be considered in terms of how they will experience it and the diversification of the forest should be focused at different levels of detail in different places, with most being concentrated around the main visitor locations and along walking trails.

There are many elements of diversity that can be built into the forest, such as different ages of trees, different species, thinner or denser stands, single or mixed species stands, multi-structure or single structure, dense undergrowth or no undergrowth etc. Other non-forest elements that should be incorporated include water

Box 45: Forest diversification

A forest diversification project in the Erdöanya-Valley in Hungary has, through deliberate forest management practices, turned a forest monoculture into a diverse aged forest which provides much better recreational and ecological opportunities.

Photo: Drexler

Box 46: Restoration of disturbed land

An area where the former oil shale workings have been reforested and are used for special recreation – "jeep safaris", Aidu mine, Estonia.

Photo: Bell

Box 47: Filtered view through forest towards light

Photo: McCormack

A shallow strip of trees acts as a screen to filter the view of open ground beyond. This creates a sense of space beyond, engendering a sense of mystery and stimulating curiosity and a desire to explore (see color plate 25).

Box 48: Varied forest structure

Variation in the forest structure should be introduced sequentially along walking routes. This can be used to stimulate movement as well as to suggest places to pause, rest and look. It can be realised by varying stem density, path width, tree height and stem visibility (through thinning and pruning).

Photo: McCormack

Box 49: Roadside design of forests

Photo: McCormack

Attention should especially be paid where forests flank roads to avoid monotony by ensuring that bends are included as well as the incorporation of open stands, recesses and clearings that allow visual penetration into the forest and through thinning / tapering out bends and edges. The eye naturally seeks to "enter and understand", especially where attracted by light, a splash of distinctive vegetation or water.

Box 50: Roadside felling

An example where the small scale of the felling, its irregular shape and the retention of some clumps of trees has reduced the visual impact and made the felling area a point of interest. North Vidzeme, Latvia.

Photo: Bell

Box 51: Dramatising the sense of passage using pinch points

Photo: McCormack

Stem spacing can deliberately be allowed to become dense at particular locations on both sides of a path or to combine with a high rock to create a shadowy narrowed pinch-point that intensifies the sense of passage giving way to one of exhilaration by release into an open and well lit space beyond. Such compression along a route can also be used to frame a view beyond.

(still and moving), wetlands, rocks, open glades, archaeological features, and views out of the forest. The degree of diversity introduced into the forest should reflect the wider landscape and it is unsustainable to create too much artificial diversity. The problem has often been that the forest has been reduced in diversity by management and this can be reinstated over time.

For recreation more diverse forests are often preferred. However, smaller areas of monoculture or culturally influenced forest types may actually add to the overall diversity of forest recreation experiences. From experience and experimentation it can be demonstrated that over a relatively short time span even monoculture forests can be changed to more diverse forest types.

Aesthetic improvement can also be achieved through reclamation projects where abandoned and degraded sites can be turned into both ecologically and recreationally improved areas through efforts at judicial afforestation. This has been widely used in the UK, northern France, eastern Germany and Estonia on former coal, brown coal or oil shale mines. Sometimes the landform is reshaped first and at other times the forest is planted on the dumped material, which can make for interesting and challenging terrain which offers exciting recreation possibilities.

Box 52: Dramatising the sense of passage at road crest between deep cut banks

Photo: Anon

A "squeeze-and-release" experience can be made even more intense where the crest of the road cuts deeply through steep banks which are surmounted by tall trees of bare stem, such as Scots pine, soaring to lofty heights. This not only intensifies the sequential experience of ascent and anticipation of a gateway passage and release, but also introduces a sense of the sublime: the exalted power of nature versus the human as minute and vulnerable (see color plate 26).

Box 53: Aesthetic effect of prospect

Views from a high viewpoint down into the forest provide a vista or, at least, a sense of command that satisfies a need for visual control. This appeals to the innate hunter instinct and stimulates an aesthetic response.

Photo: McCormack

Box 54: Well light clearings as basis for sense of place

Photo: McCormack

A clearing in a forest provides spatial relief along a forest trail. The basis of a sense of place may be engendered where the canopy is sufficiently open to flood the space with light.

Juxtaposition with the forest trail will be intensified where the ground cover comprises lush grass (meadow or lawn), so articulating the space and suggesting a pause on the journey. The location of a bench in such a space further prompts this response as well as providing a focus.

Box 55: Mysterious and mystical power of a waterfall in the forest

Photo: McCormack

The squeeze and release can be engendered by a high waterfall that issues forth from boulders between lofty conifers. The Vartry River falls dramatically from above the viewer at the Devil's Glen, County Wicklow, Ireland, creating a certain sense of mystery in regard to source as well as a mystical longing (see color plate 27).

Understanding of the possibilities for forest management to create interesting internal landscapes is important, so that these can be incorporated into regular planning of site or stand level management activities such as thinning. Clear cutting along roads is also an important area for specific consideration of landscape aesthetics.

Box 56: Art in the forest

Photo: McCormack

Adding elements of art to the forest may also enhance the aesthetic (involving the intellectual) dimension of the forest experiences. This can be particularly effective where the pieces of art interact with the forest and prompt the visitor to think about and see the forest in different ways.

The Devil's Glen Wood, County Wicklow, Ireland functions as an outdoor gallery for a Sculpture in Woodland programme (see color plate 28).

Various countries have introduced guidance for forest management with landscape quality and recreation in mind. These are often linked to sustainable forest management (SFM) (taking cultural and social aspect into account in forest management) and also certification. These guides and standards are designed to be aligned each of the criteria and indicators for SFM agreed at the Third Ministerial Conference on the Protection of Forests in Europe (Lisbon 1998).

Box 57: Nature's "art" in the forest based on human involvement

It is worth noting that the interaction of nature (e. wind) and human (forestry) forces can give rise features that can be deemed to be artistic. For example, attractive shapes can result from windblown plantations (see color plate 29).

Photo: McCormack

8.3.2.3 Design of Infrastructure

Infrastructure is needed in order to facilitate access and enjoyment of the forest while reducing the damage or conflicts caused by recreation. As a general rule, as little infrastructure as possible should be constructed so as to ensure that the forest landscape is not dominated by unnecessary construction and that the visit is to the forest, not the recreation site. Having said that, however, if infrastructure is to be installed, it should be of a high quality which looks good, performs well and is cheap to maintain as well as sustainable. The need for recreation facilities and infrastructure varies according to numbers of visitors, proximity to urban areas and the sensitivity of the site. Remote forest locations with fewer visitors usually require little infrastructure or a sensitive approach to facilities being developed otherwise they will detract from the wilderness experience.

Quality is reflected in many ways

- In good and aesthetically pleasing design.
- In solid and safe construction suited to the different user groups that is fit for purpose and yet maintains a natural feel.
- In good quality materials blending into the forest environment and in some cases using traditional materials.
- In good craftsmanship in the construction and in some cases using traditional construction techniques as part of maintaining the skills and maintaining a unique identity of the forest or region.

- In good maintenance of the facilities.
- In good connection with the user groups and a willingness to adapt/establish new facilities to meet new recreational needs.

Design Guidelines for Infrastructure

A number of guidelines on the design, planning, implementation and maintenance of recreational infrastructure and facilities are available. At the most basic, forest recreation is about walking, cycling or skiing along a trail through a forest. Such trails may be informal or make use of forest roads, requiring limited maintenance or upkeep. However, many forests offer much more and the provision may include a wider range of facilities for day visitors of tourists.

Box 58: Design of outdoor recreation

Photo: Bell

A book which provides comprehensive guidance on the design of recreation facilities in the outdoors.

The main aspects to be considered in infrastructure planning and design concern the different elements of a visit:

- The approach and entrance
- Parking
- Information
- Picnic places and toilets
- Children's play
- Trails
- Special facilities such as tourist attractions
- Interpretation and educational facilities
- Additional facilities such as shelters and towers

All facilities should, as a matter of course, consider access for all, since many members of society have difficulties of a physical or mental nature which can limit their participation in recreation unless it is specifically incorporated (older people, physically or mentally disabled, people with young children etc.).

The Approach and Entrance

The entrance to the forest from a public road marks the point at which the visitor is hosted by the forest owner. The entrance should be marked in some way,

Box 59: Entrance point in Finland

A good example of an appropriate design for an entrance information points at the entrance gate of Pyhä-Luosto National Park, Finland. Information can be given in several languages and include instructions on litter prevention, information on forest management activities, etc (see color plate 30).

Photo: Kuure

perhaps with a special feature or sign. Information at access points is a key factor in helping to give forest visitors an overview of the recreational opportunities on offer. However, the provision can vary greatly, ranging from a lack of signage and information in some situations to too many diverse signs detracting from the aesthetics of the forest. Providing clear and informative messages to visitors in an aesthetically pleasing way improves the recreational experience. A nationwide forest signage programme in Denmark illustrates how clear information, often in simple pictures, can be understood by all ages and nationalities and be aesthetically pleasing. This programme implemented at all state forests and is now available to all forest owners free of charge, with detailed guidance from a design handbook. The British Forestry Commission has a complete sign system using symbols and special signs at forest entrances and elsewhere around the site.

Box 60: Entrance point in Latvia

This entrance point is marked by a gateway and a sign big enough to be read from the public road. The gateway is based on traditional Livonian fortress construction (see color plate 31).

Photo: Bell

Parking, Public Transport

Resorts and Parks must also provide a well designed car parking that is sensitively located in order to avoid excessive disturbances to the forest environment. Providing public transport or shuttle systems from the nearest village to the forest are an alternative way of reducing the impact form visitors and the amount of infrastructure needed in the forest. Normally, car parks are made of materials that are found locally, avoiding urban materials where possible. Ideally, they should be laid out under the trees which provide screening and shade. Places set aside for disabled users should also be included.

Box 61: Car park in Scotland

A small car park set out amongst the forest at Glen Affric in Scotland uses local materials on the surfacing, and is divided into bays to reduce the impact of the cars in the view (see color plate 32).

Photo: Bell

Box 62: Car park in France

An example of a larger car park set out under the widely spaced pine trees near the coast at Cap Ferret in Aquitaine, France. The logs and posts keep cars confined to surfaced area and prevent erosion (see color plate 33).

Photo: Bell

Information

Visitors generally need information at the point of arrival. Such information should show the facilities, explain the main aspects of the area, display special information such as events or activities, and display any rules or regulations. It may be necessary to provide the information in several languages and the use of symbols reduces the need for multi-lingual text and saves on space.

Also brochures and maps are important means of communication. In Denmark, there is a program where maps and brochures are produced for most forests and provided in a small box at the entrances. So visitors can pick a free brochure with a map which allows them to get an overview of the forest, different trails, facilities, a short description of the history, special features and experiences as well as information on things to be careful of.

Box 63: Information sign

An example of a well-designed information sign giving visitors details of a trail network. Rosehall community forest, Scotland (see color plate 34).

Photo: Bell

Picnicking and Toilets

One of the most popular activities to take place in forests is a picnic, with or without a fireplace and grill. These may be equipped with tables, benches, stone rings for fires and litter bins. In most of Europe it is normal to provide places for groups of people to have a picnic away from other people, so distributing fireplaces and benches around an area, possibly screened by trees or bushes.

Box 64: Picnic site in Switzerland

This picnic site in Allschwil Forest in Switzerland meets the visitor preferences of the area.

Photo: Hegetschweiler

Box 65: Picnic site in Cyprus

This picnic site in the Troodos mountains of Cyprus is very busy in summer as people escape the heat of the lowlands and have a large sociable barbecue (see color plate 35).

Photo: Kazafaniotis

Box 66: Toilets facilities in Scotland

Photo: Bell

This is a good example of a composting toilet built from local materials and fully accessible to disabled people at Abriachan community forest, Scotland (see color plate 36).

Box 67: Toilet facility in Cyprus

The toilet facilities in Pykni Forest in Cyprus are an example of maintaining a traditional building style that provides safe and modern sanitary facilities at recreational sites, and it s also fully accessible to disabled people (see color plate 37).

Photo: Kazafaniotis

In southern Europe people often prefer to be more sociable and are happy to visit large picnic sites where many people can gather together.

Busy sites with a high turnaround of visitors and where visitors may have to travel quite long distances to get there may need to be provided with toilets. These can range from small and simple, of the "earth closet" or composting variety to much larger facilities with running water. All should be built using appropriate materials and sited so as to be easily found from all parts of the site. Access for disabled people should also be considered in their design.

Play Facilities for Children

Some forest have formal play facilities as part of the experience provided, whilst others are increasingly being developed to enhance informal play. Play in forested areas enhances both the physical and mental development in children more than formal, urban playgrounds. Forest playgrounds exist in many European countries and generally use natural materials such as wood, stones, plants and soil to construct play environments. Plant areas can be concentrated or part of a play trail. They need to be laid out for different ages of children and be located within the forest as part of the larger infrastructural design.

Box 68: Provision of play structures in the UK

In the UK, some work has been carried out on the artistic qualities of the play structures. The inclusion of large wooden play structures in the forest designed by an artist/sculptor is very popular among visitors and many elements have become visual symbols (see color plate 38).

Photo: Worthington

Box 69: Play forests in the Netherlands

In the Netherlands, forest playgrounds are very popular. Some of them also allow informal play where children can use forest material in a creative way. The Play Forest Mastbos is a good example of local involvement in playground creation and construction. Children were involved in the planning process and scouts took part in the construction of the facilities (see color plate 39).

Photo: Staatsbosbeheer

Important is also providing information on playgrounds. A good example is Denmark where a web based data base of currently 75 nature playgrounds has been established (http://www.skovognatur.dk/Ud/Aktiv/naturlegepladser/default.htm) which provides families with a quick overview so they can find forest playgrounds and visit new areas.

Box 70: Design guide on play Forests

Forest playgrounds have to be designed to be safe, but also challenging for children. A design guide "Design guidance for play spaces" from the UK helps forest managers through the process of creating play features in the forest; from setting up a project team, consultation, design process and principles to maintenance.

Photo: Forestry Commission

Trails

The creation of extensive trail systems throughout forest areas improves access. The Skaneleden trail in Scania, Sweden has been gradually expanded through scenic areas (including forests) since it was first established in 1978; it now includes over 1,000 km of trail and is used on average by 100,000 people each year. Many forests contain networks of forest roads and tracks which can be used as trails when they are not being used for forest management, by visitors on foot, cycle, horseback or skis. In other circumstances new trails are developed which take visitors to see especially attractive elements such as viewpoints, special nature conservation areas or waterfalls, lakes, wetlands etc. In busy areas it may be necessary to separate trails between users such as horses and cycles because of possible conflicts. In places where the site is vulnerable to wear and tear it may be necessary to limit access to trails and to construct special surfaces in order to protect the site. The forest should be managed to provide the best internal landscape along trails (see earlier section in this chapter).

Box 71: Recreational trail along an abandoned train line

An abandoned train line in a forested valley in Sabor Ecopista, Portugal has been converted into a recreational trail for walkers and cyclists. Whilst each of these examples have been linked to recreational facilities within the forest they have also become tourist attractions in their own right (see color plate 40).

Photo: Castro

Box 72: Trails in wetland areas

This boardwalk allows access to and across a wetland nature reserve at Ķemeri National Park in Latvia. It uses local materials and protects the fragile landscape (see color plate 41).

Photo: Bell

Box 73: Trail for disabled use

Photo: Kaae

Trails for wheelchairs need to be well-surfaced, wide and level. This is an example from Dyrehaven Forest in Denmark (see color plate 42).

Box 74: Forest access for disabled use

In Lithuania, as part of the 2003 Disabled Year, a large project was developed to aid forest access for all groups. Importantly representatives from disabled groups were involved in the project, which involved 42 forest enterprises establishing 250 facilities (see color plate 43).

Photo: Reseiniai Forest Enterprise

Steep gradients, soft surfaces and steps reduce access for many disabled and (baby) push-chairs users or elderly visitors. Careful planning and design are required to give access to as many people as possible. Awareness of providing all-ability access and signage-for-all is increasing, with many forests now having trails able to accommodate wheelchair users and people with other types of disabilities. Some trails have been carefully designed to provide different sensory experiences for all.

Good examples of trails with all-ability access have been provided in many countries such as the Kambous tou Livadiou nature trail in Cyprus. Several trails also provide good information in braille, and sensory boxes can stimulate experiences of nature through the sense of touch. The Sentiero Natura per diversamente abili and the trail for all senses – campigna di Santa Sofia in itlay combine accessibility with interpretation. Ryckevelde and Zonienwoud from Flanders are examples of trails accessible to all, with interpretation in braille and sensory elements.

Box 75: Interpretative design for trails

This photograph showing interpretive design from "The trail for all senses", Campigna di Santa Sofia, Italy illustrates how tactile panels can be created that are of interest to all (see color plate 44).

Photo: Giovanni

Link to Tourism Facilities

Existing forest infrastructure that might historically have been used for forest management can be al link to tourism facilities. Kemence Forest Railway in Hungary, and the forest railways at Białowieza in Poland and the Vydro Valley in Slovakia are examples of the conversion of old forestry railways into tourist transport systems.

Box 76: Access to forests by railway

The 30 km Porboly – Baranyfok railway line passes through the Gemenc nature reserve of the Danube-Drava National Park. The nature reserve can only be visited by railway.

Photo: Laczkó

Interpretation and Educational Facilities

The recreational experiences in forests may be enhanced through interpretation, visitor centres, educational programmes and tours, providing more in-depth understanding of the forests. Media and information technologies, modern materials, innovative design are the new and innovative elements that can improve the enjoyment of any activity in the forest, from education to sport. Interpretation and guiding themes can vary but should be broad in respect to forest functions and ecological issues and/or more focused on specific topics and activities of the local forest. They can also include cultural or historical aspects of forests and of the lifestyle of societies linked to it.

Most national parks and larger forest protected areas provide visitor centres and a number of approaches and styles are found across Europe. Some examples from different regions of Europe are presented below and represent good practices, reflecting the emerging features highlighted, the quality and completeness of information, the peculiar styles and characteristics and the innovative awareness/communication techniques.

In Finland visitor centres create a customer service network that covers the whole country and the most significant protected areas. There are 22 visitor and nature centres and four other customer service points in Finland. Visitor and nature centres are most useful places to visit prior to an excursion. Visitors can obtain information on their destination as well as rest before setting off on a hike. Intriguing exhibitions and impressive AV presentations bring Finland's most spectacular sights to visitors' fingertips. Moreover, there are other smaller information points, for example nature information huts, where useful information can be found. All visitor centres, nature centres and some outdoor information points offer tours designed for groups. They all are coordinated and managed by Metsähallitus (www.outdoors.fi).

Box 77: Visitor centre stimulating curiosity and imagination

One of the ten visitor centres is the National Park "Foreste Casentinesi, Monte Falterona e Campigna" (Italy). The visitor centres stimulate the curiosity, interest and imagination of tourists and visitors while at the same time providing knowledge and instilling a respect for nature The use of movies, slides, plastic models, suggestions and educational games are all aimed at establishing deeper connection between the visitor and the cultural tradition and nature of the protected area (see color plate 45).

Photo: www.parks.it/parco.nazionale.for.casentinesi/

Box 78: Re-using existing facilities as visitor centre

The Visitor centre of the Athalassa National Forest Park (Cyprus) makes good use of old buildings with cultural and architectural value. The centre is equipped with modern facilities such as laboratories with micro-scopes and stereoscopes, computer information screens, dioramas, a film/lecture theatre and exhibitions of geology, fauna and flora. In the garden there is a small amphitheatre where lectures to students and organised groups are given (see color plate 46).

Photo: Kazafaniotis

An example of a large and coordinated effort is the visitor centers in Foreste Casentinesi, Monte Falterona e Campigna National Park in Italy. This Park has ten visitor centres, each one with a permanent thematic exhibition on major issues of the Park, with visual displays and innovative communication tools. Furthermore, the centres are attractively designed and furnished with natural materials coming from the local forests.

Mobile visitor centres provide a flexible solution where the information can be provided in several locations in a rotation or where the visitor flux is strictly related to the most pleasant season. An example from the UK shows the conversion of a steel container into a mobile visitor centre which is robust, secure and mobile on a lorry.

Box 79: Visitor centre with nice architecture

The visitor centre at Glendalough (Ireland) comprises a cluster of neatly formed blocks, each housing specific visitor functions, such as reception, display, auditorium and services. The architectural language expressed in form and materials is appropriate to the naturalistic setting. Visitor facilities include a well designed car park, picnic areas, toilets and board walks that meander across a marsh (see color plate 47).

Photo: Pöhlmann

Huts and Shelters

Some facilities are climate-dependent. Shelters are a benefit in both hot, sunny climates as well as in the very cold, wet regions. Access to drinking water is often a necessary requirement of hot Mediterranean climates. Shelters allow people to rest, to extend their stay or to engage in other activities. Materials and method of construction affect the appearance and appropriateness of recreational facilities. Features that reflect local, traditional building methods and materials connect with the surrounding countryside environment, but must also be accessible to all, be easy to clean and maintain and provide a high standard of hygiene.

Box 80: Visitor centre with focus on animals

The enlargement area of the national park "Bayrischer Wald" (Germany) combines recreation and tourism with conservation goals. A central visitor centre "House of the Wilderness" is linked to two game reserves with Przewalski horses and indigenous cattle. Also wolf and lynx reserves are located directly next to the "House of the Wilderness" which also provides a cafeteria, exhibition space, sale of regional products, toilets, assembly rooms and a view to the wolf reserve.

Photo: Pöhlmann

Box 81: Mobile steel visitor centre

This conversion of a steel container provides a secure, robust, mobile visitor centre that can move from site to site on the back of a lorry.

Often buildings in remote forest locations are vulnerable to vandalism. Buildings are costly to erect and staff, particularly where visitors might be seasonal. In some sensitive areas there are planning restrictions preventing development (see color plate 48).

Photo: Worthington

Box 82: Picnic, shelter and fire place

This picnic, shelter and fire place in the Arctic Circle Hiking Area in Finland is an example of an attractive design, accessible to all with minimal risk of fire (see color plate 49).

Photo: Paso

Box 83: Shelter in Scotland

This shelter also acts as an information centre. It is made of big logs with a turf roof. Rosehall community forest, Scotland (see color plate 50).

Photo: Bell

Box 84: Picnic shelter in Finland

This shelter from Evo, Finland is big enough for a large group of people to visit and have a picnic, cook food and spend a good time (see color plate 51).

Photo: Bell

Box 85: Historic shelter

The historic infrastructure in Budakeszi Wildlife Park in Hungary gives the park a special identity (see color plate 52).

Viewing Towers

In some places where the landscape is quite flat and the forest rather dominant, it is difficult to obtain views, so towers can be constructed so as to be taller than the trees.

Box 86: Viewing tower

This tower, built of good-sized logs, provides wide views over the forest and wetland in this part of Estonia where panoramic views are otherwise unobtainable.

Photo: Bell

8.3.3 Unique Experiences

Birgit Elands[5] *and Berit C. Kaae*

With contributions from: Olgirda Belova, Maria Bihunova, Art McCormack, Dóra Drexler, Joel Erkkonen, Lubica Feriancova, Savvas Kazafaniotes, Gudrun Van Langenhove, Barbara Mariotti, George Pattichis, Pieter Roovers, Dan Rydberg.

8.3.3.1 Introduction

As society changes, outdoor recreation preferences change as well. Social, economic and demographic changes, increasing awareness of the need to protect the environment as well as innovative technologies and facilities all contribute to new recreation experiences. The natural environment is on the one hand appreciated for its ecological values and its benefits to human health and spirituality. On the other hand the natural environment is increasingly seen as a source of consumption for amusement and adventure.

In urbanised societies, where people need to escape the regularity, hardness and busyness of the built environment, nature is considered as a place where one can calm down and express oneself. Outdoor recreation is increasingly becoming an expression of lifestyle and identity. The way people express themselves, however, is as varied as people can be. Consequently, unique recreation experiences vary from *fast recreation* (adventure, fun, novelty) to *slow recreation* (environmental responsible behaviour, contemplation, and familiarity). To experience nature becomes more important to visitors while at the same time visitor's experiences become more diverse.

For forest managers, recreation needs to be more than the provision of access, infrastructure and facilities for a range of activities. Forest managers should be aware of the fact that a certain kind of activity can be experienced in different ways: one can hike for merely social reasons, e.g. to be together with your family, or hike to find solitude and quietness within yourself (Elands and Lengkeek 2000). Consequently, foresters need to take the recreation experience into consideration for planning and design. This implies that a forest should provide points, spaces or zones in order to experience nature in different ways.

Through comprehensive planning at local and regional level, discrete zones for one or a number of recreation experiences could be included. Quite often, discrete zones are classified according to the intensity of human interference: varying from very intensively used zones with all kinds of user faciltiies and infrastructure and exhaustive forest management until extensively zones with hardly any facilities and infrastructure and very limited forest management. A good example of this type of planning provides the Recreation Opportunity Spectrum (Clark and Stankey 1997),

[5]Wageningen University & Research Centre, Forest & Nature Conservation Policy Group, Droevendaalsesteeg 3, 6708 PB Wageningen, The Netherlands, e-mail: birgit.elands@wur.nl

which in this case can be framed as Recreation Experience Spectrum. In Denmark, a method called "experience mapping" (Caspersen and Olafsson 2006) is increasingly used to identify different types of recreational experiences found in forests and other nature areas. This method can be used in planning to ensure that recreational experiences are differentiated and not disturbed (e.g. areas of tranquil forest experiences are not disturbed by over-equipping the forest).

Principles and examples of providing unique experience in forests and nature areas will be illustrated in the following sections:

- Adventure
- Imagination
- Health
- Environmental education
- Spirituality.

8.3.3.2 Adventure

The main attraction of adventure recreation and tourism is an outdoor activity that relies on characteristics of the natural environment, often requires special equipment, and is exciting for its participants (Buckley 2007). It includes activities such as climbing, caving, abseiling, kayaking, rafting, skiing, snow boarding, parapenting, horse riding, mountain biking, snowmobiling and orienteering. Participants may bring the equipment themselves, but the necessary facilities can also offered by landowners and commercial organisations, such as tour operators.

Facilities for adventure recreation and tourism can be realised and organised by forest owners e.g. through *canopy trails*, *tree climbing facilities*, and *rope courses*. A good practice example of this is the Val di Vara Adventure Park in Italy. It comprises primarily treetop walks/canopy walks with an extensive system of adult and children paths along with equipment and trainers.

Another way of organising adventure in forest and nature areas is by means of organising long distance trips, during the night or daytime, in difficult circumstances, and with the help of obstacles and puzzles. All these aspects should challenge a person's physical and psychological endurance.

Some examples are:

Orienteering. This is a group of sports that require navigational skills using a map and compass to travel through mostly natural and unfamiliar terrain. Most common activities are foot, mountain bike, ski and trail orienteering, Mostly, a type of competition is involved.[6]

[6]http://en.wikipedia.org/wiki/Orienteering

Box 87: Val di Vara Adventure Park

The Adventure Park of Val di Vara (Italy) includes suspended trails integrated in the forest canopy, thanks to cable installation, wooden platforms, and ropes allowing the users to move safely on canopy trails where balance and self concentration (even more than physical dexterity and condition) are essential. All the trails require personal safety equipment. The construction technique employed for each trail is tailored according to its level of challenge. There are special trails for children (up to 5 years) as well as easy trails right up to those that are extremely challenging (see color plate 53).

Photo: Mariotti

Geocaching is a modern form of treasure search. People search "hidden treasures" with the help of a GPS device and leave them in a new hiding place. The coordinates are exchanged on geo-caching websites.
Wilderness survival trips, in which living in the wilderness as a way of life is aimed at. *Woodcraft* and *bushcraft* are typical forest related examples of wilderness survival experiences in which the acquisition of ancient skills and traditional

knowledge with respect to living in the woods (e.g. hunting, fishing, and camping) is being trained.[7]

8.3.3.3 Imagination

Imagination is essential to leisure time: for a moment we escape our daily duties and responsibilities and we participate in an experience. In our imagination, we can travel all over the world and discover new places and cultures; we can live in another era and play that we are in ancient Rome or colonial England; we can discover different personalities through imagining that we are Jeanne d'Arc or John Lennon, and we can create an illusion of something which is neither present nor exists and thus builds a world of phantasy (Ricoeur 1994). Imagination is a change in space, time, culture and identity or a merge into non-existing reality.

Imagination plays an important role in recreation and tourism. Sensory experiences are important: sight, hearing, taste, smell, touch, but also through our movements balance and acceleration, temperature and kinetic sense.

How can forest managers facilitate imagination? They can facilitate imagination through:

- the creation of specific places that link to existing places, times, cultures and personalities (e.g. *open air museum*)
- the creation of specific places that refer to a phantasy world (e.g. *fairy tale forest and dwarf wood*)
- the organisation of participation in existing or phantasy worlds (e.g. *live role play*)
- the use of techniques that stimulate our senses (e.g. *sensory garden*).

Good examples for an open air museum, fairy tale forest, live role play and sensory garden will be described below.

A good example of the creation of a specific place in a specific time period is the *Open Air Museum* in Gyftocampos, which is part of the Greek Zagori area of Epirus. This open air museum exhibits the traditional lifestyle of mountainous people. It is a popular cultural tourism attraction.

A good example of the creation of a specific place that refers to a phantasy world is the *Fairy Tale Forest and Dwarf wood* of Tervete Country Park in Latvia. The forest is used for carvings of fairy tale and dwarf characters and houses. It is a very popular place for familites with kids and those people who are still interested in fairy tales, dwarfes and other phantasy images.

[7] http://en.wikipedia.org/wiki/Woodcraft and http://en.wikipedia.org/wiki/Bushcraft

Box 88: Open Air Museum of Gyftocampos

The Open Air Museum of Gyftocampos in the Zagori area of Epirus shows the traditional lifestyle of mountainous people. It includes a reconstruction of a community of nomad sheep raisers, the Sarakatsani. The Sarakatsani still meet here on the first Sunday on August each year. The celebration goes on for three days. The ritual contains strong elements of a community of sheep raisers that was secluded until recently but which experts believe has lived here since Neolithic times (see color plate 54).

Photo: Aristotelian University of Thessaloniki Department of Forestry and Natural Environment

One example to merge in a phantasy world is *Live Role Play* which is a popular new recreation activity in many Danish forests. It involves large groups of children and teenagers playing and dressing up to perform different roles (elves, trolls, witches, etc.) and using a scripts to carry out the quest (somewhat like computer games but, instead, carried out in real life in forest settings). Up to

800 children/adults are participating in the live role play at one time in the forest. There are permits granted per forest district, thus ensuring that the events are spread out in time and space so as not to disturb the breeding seasons of animals etc.

Box 89: Fairy Tale forest and Dwarf wood, Tervete Country Park

The Fairy Tale Forest and the Dwarf Wood are inspired by the fairy tales and plays by the Lativan writer Anna Brigadere (1861—1933). Magically sculptures of fairy tales characters and members of the friendly dwarf family inhabit Tērvete Country Park making this a wonderful place to visit and photograph. In the summer period, visitors can meet some characters alive performed by actors. Visitor can also walk through a unique pine forest where most trees are older than 300 years. The park also has a playground with swings, shelters, tables and benches. (www.mammadaba.lv/pages/top_detail.php?l=en&id_object=61) (see color plate 55)

Photo: Latvijas valsts meži

Box 90: Live role play

Live role play is a type of theatre where participants play different roles but where the play develops without a fixed manuscript. It may enact different time periods such as the middle age, the Viking era, an adventure world of trolls, elves and other creatures. Both children and adults participate in this activity. The groups of people participating can be quite large and often involve several hundred at a time. Many participate on occasion but at least 80 role play clubs have been established in Denmark and the popularity of the activity is growing. In the summer regional and national summer camps of role players are held lasting up to one week. Some more commercial team-building companies also use live role play as part of their activities (see color plate 56).

Photo: Andersen

An example in which our senses, especially our tactile sense, are stimulated to experience the natural landscape differently is the *Bare Feet Path* in *Opende* (Netherlands).

Box 91: Bare feet path

The Bare Feet path or in Dutch dialect "'t Blôde Fuottenpaad" is situated in a nature area in the northern part of the Netherlands. One can feel, smell, see and hear the surrounding landscape. The path is 4 km and includes walking through meadows and water, on sandy paths en stones. People can walk, sit and relax, and climb. Simple facilities, such as the game "bare feet dance" and "hammocks", make the hike pleasant (www.blotevoetenpad.nl) (see color plate 57).

Photo: Buro Kloeg, the Netherlands

8.3.3.4 Health

It is widely demonstrated that nature provides benefits to human health and well being, oriented to physical as well as to psychological disturbances due to urban living conditions and contemporary lifestyle, Thus forests are increasingly being used for both preventive measures, specific therapies and treatment. There is a broad range of literature about forests and health involving aesthetics, species composition, landscape etc. in relation to the treatment of specific diseases.[8]

This section will illustrate two aspects of health benefits:

- Fitness and physical health
- Mental health

[8] A COST action dedicated to this aspect is still running: COST E39 "Forests, Trees and Human Health & Well-being". More information can be found at: www.e39.ee.

Fitness and Physical Health

Physical health can be stimulated through a variety of facilities and infrastructure. Two typical examples will be illustrated: Nordic Walking and Fitness parcours. Preventive measures include trim trails and exercise equipment in the forest which helps people stay in shape and reduce their risk of cardiovascular diseases, obesity and other health problems.

A good example of this is the increasingly popular sport of *Nordic walking*, where people simply bring their own walking poles and use existing trails and facilities. In some European countries, as e.g. in The Netherlands, a convenant on Nordic

Box 92: Nordic walking

Nordic walking was first launched in Finland in 1997. It has rapidly increased in popularity and today nordic walking is a well-known fitness sport around the world. Nordic walking is part of a year-round Nordic Fitness Sports™ concept developed by Exel, which offers fitness enthusiasts an easy and fun way to enjoy a healthy and active lifestyle.

The Nordic Fitness Sports Park™ Pyhä-Luosto was opened in June 2007. It offers a 100 km network of routes for nordic walking, mountain biking and snow shoeing in the surroundings of Pyhä-Luosto tourism destination and the Pyhä-Luosto National Park, comprising 13 different routes of varying levels of difficulty. All the routes are clearly marked and signposted to ensure easy orienteering in the park (see color plate 58).

Photo: Erkkonen

Walking helped integrate this new activity into the forest. The covenant of various institutions has established a common strategy to accommodate Nordic walking as an activity and includes design and management of NW paths, special activities and commercial user fees, rules of conduct, instructors training, publicity and promotion. Nordic Walking can be done in groups and appeals to all age groups. However, older age groups in particular find it appealing as it combines less strenuous physical exersice with social interaction. While it requires very little more than basic trails, the provision of car parking and restroom facilities are desirable.

The second example deals with a *fitness track* or *trim trail*. Two good examples of preventive sport activity come from Cyprus and Lithuania. The trim trail in Athalassa National Forest Park in Cyprus provides diverse opportunities for outdoor exercise and is connected with the trail network of the Park. Another example is the wellness pathway in Kursiu Nerija National Park in Lithuania which is equipped

Box 93: Trim trail at Athalassa National Forest Park

The Trim trail is a part of the recreational activities offered by the Forestry Department at Athalassa National Forest Park (Cyprus). The general objective of this trail is to give the opportunity of outdoor exercise to the general public. It is connected with the trail network of the Park. The trail has several exercise units such as jumping hurdles, parallel bars, flexibility rings, horizontal ladder, chinning bars, balance beam, sit up platform, step up logs, straddle jump bar etc. The trail provides fun and fitness for all levels of physical ability (see color plate 59).

Photo: Kazafaniotis

with special woody facilities for physical exercise. Furthermore, along the path 70 woody mythological statuaries provides a special atmosphere and character to the facility (see www.forest-recreation.info).

Mental Health

In addition to the preventive functions, forests are also being used to treat diseases such as stress, eating disorders etc. The therapeutic uses of forest are undertaken by professional institutions using the unique qualities of forest environments as part of the treatment. An example is the use of forests in Danish programs for the treatment

Box 94: Therapeutic use of the forest

The therapeutic functions of forests are increasingly being recognized. In Denmark, forests are now used for the treatment of eating disorders among young adults. The treatment program involves single groups of young women (6–8 persons/group, 20–30 years old) with Anorexia Nervosa, Bulimia Nervosa and Bing-Eating Disorder. Objectives are to strengthen the participant by building self-esteem, enhance trust in others, try new foods outside familiar and safe settings, gain backcountry camping skills, and incorporate learned skills into daily life. It uses modifications of traditional adventure therapy with light to moderate rather than rigorous activities. Activities can include paddling, fishing, team building, trust and problem-solving initiatives, therapeutic journaling, cookouts, backpacking, solos and camping (see color plate 60).

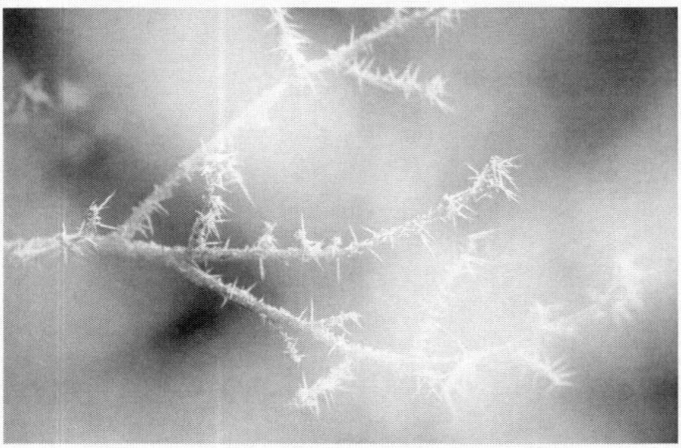

Photo: Kaae

of different eating disorders. The forest here provides non-demanding surroundings which gives the patients affected by eating disorders a "free space" from the many controls they feel trapped by in their daily lives. The forest and the therapeutic uses help the patients gain self confidence and to overcome their illness. Also stress is positively reduced by forests. Therapeutic use of forests is wide-spread in other parts of the world including the U.S.

8.3.3.5 Environmental Education

In urbanizing societies contact with nature is not self-evidently anymore. Especially young people often have limited experiences with the natural environment. Politicians and natural resource managers might fear that in the nearby future public support for protecting nature might decrease (Elands and Van Koppen 2007). In order to prevent this, a positive and strong attachment of (urban) citizens to forests and other types of nature areas needs to be supported. Besides, it is not always obvious that users once visiting a forest behave in a responsible way, both in respect to the environment and to other visitors. Consequently, enhancing environmental awareness has become an important issue in forest management.

Basically, there are two ways forest managers can contribute to environmental education:

- Through the supply of facilities aimed at education and interpretation;
- Providing a setting for outdoor education and nurseries.

Supply of Facilities Aimed at Education and Interpretation

Environmental education can stimulated through the supply of certain facilities. We will deal with visitor centres, natural trails, forest heritage parks, wildlife observation and educational hiking tours.

The most obvious example is a *visitor centre* (see also Section 8.3.2) of which information services, interpretation and education are the most important functions. Within or outside the visitor centres a range of different interpretative facilities and techniques are used. New technologies allow the provision of new and innovative tools to enhance the outdoor experience; displays, boards with photo and text, hand-on displays, audio-visuals shows, computer animations, web-cams of specific feature, question boxes, observation platforms and many more techniques are used to communicate the interpretative messages to the visitors in forests. However, such a variety can also become too much at times and overwhelm the visitor if not judiciously selected. Thus a balance has to be established according to the location, the themes, the purpose and the final users. Awareness of the quality of the information is also important, as is knowledge of the receivers, including children, people with different disabilities etc.

A second example is *nature trails*, which are paths with natural features labelled especially for educational purposes. Sensory experiences can also be stimulated through the use of specific materials and techniques. A good example of providing sensory experiences on a nature trail is the path for visitors with visual disability in the Zonienwoud in Belgium.

Box 95: Smell and feeling boxes with info boards in serigraphy and braille

Zonienwoud (Flanders, Belgium) also provides a path for visitors with visual disability. A tour of 1,500 m is provided by guides trained especially for visual disabled visitors. They let visitors feel and smell the forest (trees, trunks, flowers, etc.). Special features: "voelbakken" and info boards in braille (see color plate 61).

Photo: Raes

Thirdly, forest *heritage parks* can be created. Forests invariably contain a range of cultural heritage features including a history of traditional forest management; this can be of interest to recreational user groups and interpretation can help them understand the traditions of forest environments. Also the marking of natural heritage, such as glacial remains, can be of interest to forest visitors.

A good example of a heritage park is the Forestry Open Air museum in Vydrovo Valley, Slovakia. It is an excellent example of an educational and recreational trail. It offers information about forestry, foresters, nature and the relations between them.

Box 96: Forestry Open Air museum in Vydrovo Valley

A location where forestry practice is interpreted for visitors is the Forestry Open Air museum in the Vydrovo Valley in Slovakia. It includes a museum on historical forestry practice and a trail system with information boards and wood art. As part of this, the old forestry train has been restored and provides recreational access on old railway tracks.

Photo: Bihunova

Another good example of forestry heritage is Dzukija National Park in Lithuania, which is an ethnographic forest village that is maintained through local residents and integrated into recreational and tourism programmes. In Portugal, forests are traditionally the providers of water and contain many watermills that in the past have been abandoned. Some areas have reconstructed these traditional water mills for tourism purposes and include overnight accommodation and information on the cultural heritage and function of the historic water mills. Devils´ Glen Wood in Ireland is a woodland that encompasses the historic and cultural heritage of the area. Furthermore, it focuses on contemporary culture through an outdoor gallery of sculpture. An education and outreach programme is connected with the facility and publications etc. are available to visitors.

In addition to cultural heritage, elements such as the geological history of forests can become attractive for recreation and nature tourists. An example is the Petrified Forest of Lesvos National Monument in Greece where the unique geological history of forests (the 20 million year old Sequoiadendron gigantean) has become an attractive and scientifically interesting site visited by thousands of tourists each year.

Fourthly, *wildlife observation facilities* make it possible for visitors to observe the wildlife without disturbing the animals and their natural behaviour. This makes it important to know the animal habitats within the forest and place access infrastructure, observation towers, and other facilities in locations that leave zones for the animals undisturbed. Typically forest locations and facilities can be created to enable wildlife observation and a few facilities such as observation towers, wildlife hides, etc. can enhance the visitor's experience.

Finally, *educational tours* can be organised. The Office National des Forêts (ONF), in France, established the Retrouvance® experience, which consists of a one week guided hiking trek in remote areas with old forest tradition and history and involves local stakeholders in a sustainable development process. Retrouvance® tours open the way for people to discover nature and environmental heritage in wild forest and mountain areas located in great tourist resorts, combining outdoor activities with the development of local economy and the preservation of the forests environment. More information about each example can be found at www.forest-recreation.info.

Nature interpretation programs are not only targeted at school children and other organised groups, but also the broader public. "Naturnet.dk" is a good example of an Internet-based way to book guided tours in the Danish forest and nature areas. A nature interpreter will leads the tour' providing interpretation during the tour. The themes vary widely and can be searched on the website. Location and time of the tour is provided and tours are generally free of charge or for a minor fee. In some tourist areas, tours are also provided in foreign languages (www.naturnet.dk).

Forests as a Setting for Outdoor Education

Forest managers do not always have to create facilities themselves. They can also offer the forest as a place for societal groups to organise educational activities, such as schools, nurseries, and Scouts. Many of the experiences with forests learned by previous generations are difficult to pass on to the younger generation as many people now live in cities. Learning about forests and nature at a young age provides a higher awareness of and interest in forest and nature in later life. Educational programmes for children are a tool to promote the future sustainable development. The internet can be used as a tool.

Forest as a setting for outdoor education experiences can be stimulated in different ways. We will deal with forest schools, outdoor classrooms, nurseries, and educational programmes.

In a number of countries, education of children on forests is part of the school curriculum. Most countries in Europe have their own education programs, quite often *forest schools* called. The forest schools in Hungary have a long tradition but now primarily focus on environmental education. It is an integrated part of the regular school curriculum – these schools have educational facilities in the forests.

Box 97: Teaching school children about forests

The Forest School movement (Hungary) provides an educational opportunity within the school's local curricula. A five-day programme is held during the school term, and the Forest School is situated in the neighbourhood of the organising school, in a natural environment. A key goal is to develop healthy lifestyle skills in harmony with nature and foster a community-minded sensibility. The learning ground is the local community and its landscape, and the local strengths and abilities of the community are utilised. Forest Schools can be established either through individual schools or the Forest School Service which supports the implementation of organised educational and learning activities by providing a proper place, vocational programmes, specialists, funds, and materials (see color plate 62).

Photo: Erdeiiskola

Some forests provide special facilities for forest education. The *outdoor classroom* at Balrath Woodland in Ireland is an example. The outdoor classroom is linked to educational information for teachers at a website. It has been developed with community involvement which may enhance the knowledge of and use of the educational facility. Besides classrooms, specially designed path along key features are used for educational purposes as seen in the Seire Cognitive path, Zemaitija National Park in Lithuania. The path provides ecological education in the national park.

Educational programmes for children are a tool to promote the future sustainable development. In Sweden "Skogen I skolan" is a national cooperative program between schools and the Swedish forestry sector to educate children on forest functions. In Denmark a similar "Skoven I skolen" forest education programme is found. Another example is the environmental education in Taxiarchis forest, Haldiki Greece. This is one of several forest education projects under way in Greece in order to make people more sensitive to environmental problems.

Box 98: Outdoor in all weather nurseries

The basic idea is that personnel and children at day nurseries allocate the majority of all activities to outdoors in nature. The educational philosophy of Outdoor in all Weather has been established all over Sweden since 1985. The intensions are to teach the children how to perform in nature and to be careful with living beings and to improve the children's individual development. Studies have shown that children from Outdoor in all Weather day nurseries are less sick, had better powers of concentration, as well as motor activities than did children from usual city day nurseries. The studies stress the importance of nature in making children's play imaginative and varied, which gives children opportunities for relating to the environment in different ways (see color plate 63).

Photo: Rydberg

Environmental education for children can be developed by stimulating the creative talents of the pupils. "*Noi e il bosco*" (The forest and us) is an Italian project based on this concept and realized by a partnership of various institutions which included a theatre company. The idea was to develop, by observing and describing the natural environment, expression abilities and creative talents of children in order stimulate consciousness of actions, choices and daily behaviours related to the respect of the environment. During the visits forests interactive performance' where educators pretended to be natural elements (trees, wind, animals, etc.) were organised in order to stimulate subjects for discussion among the students, the teachers and the educators.

A lot of excellent examples of educational programme can be found across Europe. Some good examples, information and literature to support existing national projects and to inspire the establishment of new ones can be found in the website of the European network "Learning about Forests" (LEAF; www.leaf-international.org). The project works as an internet-based network between participating countries. The network helps participants to benefit from each other's experiences and materials and to share new ideas. Over ten countries across Europe are currently actively involved. LEAF programme is relatively new in forest education and it builds on the accumulated experience of well-established programmes in the European countries.

Forest Nurseries, especially in Skandinavian countries, are located in or bring the children to the forest. An example is the "Outdoor in All Weather nurseries" from Sweden, where the personnel and children at the nurseries allocate most of their activities to the outdoors. As a consequence, the children are less sick, have better powers of concentration, as well as motoric skills than children in city facilities.

8.3.3.6 Spirituality

In urbanised societies, through processes such as individualisation and secularisation, people are disconnected from traditional spiritual places such as churches, synagogues and mosques. Forest and nature areas serve as new places for engaging in existential experiences.

In this section we introduce two ways in which forests can contribute to existential experiences:

- Forests as natural burial grounds
- Forests as areas for contemplation, prayer and meditation.

Forests can function as *natural burial grounds*, which is a green burial ground or an eco-cemetery, where the body is given back to earth. It is an environmentally sustainable alternative to existing funeral practices. The first woodland burial was created at Carlisle Cemetery in the United Kingdom in 1993. Since that time more than 200 natural burial sites have been created in the UK. Natural burial is a personal choice of people who are mindful of the cyclical nature of life and who seek to minimise their impact on earth (www.naturalburial.coop). Also Denmark now allows that ashes from dead persons are buried in the forest. So people without religious ties to the church have an other option for burials than the cemetery.

A natural burial can be seen as a conservation tool to create, restore and protect urban green spaces. Mostly, grave markers are used that do not disturb the landscape. Examples are: grasses, wildflowers, shrubs, trees, a flat stone or a centralized memorial element within the forest that provides an orientation and rest point for visitors (www.naturalburial.coop; www.eternalforest.org).

Box 99: Faith woodlands

An area of the Forestry Commission, Maulden Woods, between Bedford and Luton in England has been developed by all faith communities as a permanent spiritual and educational space. Whilst paths, benches and explanatory signs have been created along with a labyrinth through the trees no religious symbols have been installed. A natural clearing with an oak tree at the centre spot provides a focal point for reflection, contemplation, meetings and picnics.

Everyone is welcomed to discover the peace and pleasure of relaxing and thinking in the Woods, and enjoying interesting, healthy walks away from the stress of modern life.

Photo: Forestry Commission England

Forests can also serve as *areas for contemplation, meditation, and reflection*. A good example of this is the Faith Woodlands project,[9]which is shared between Luton and Bedford Council of Faiths, aims to promote mutual understanding between people of all faiths, through the common ground of the natural world. Faith Woodlands are dedicated areas of woodlands, that serve as spiritual and meditative space for a diversity of faith communities, such as Muslim, Christian, Hindu and Sikh people. The first Faith Woodland was established in Maulden Woods. Focusing on youth and women in particular, this project focuses on attitudes around sense of place, identity, belonging, engagement and participation in multi-faith, multi-cultural Britain, and thereby sharing the importance of co-existence as British citizens, of different faiths or of no religious commitment at all. The project has made an important contribution towards social harmony in a diversely populated area and introduced groups who often feel excluded from woodlands to the health benefits of exercise in nature.

Another good example is the Dutch "Nature College Foundation". This organisation emphasizes the fact that human beings are born of this earth and thus are intrinsically linked to nature and natural processes. The foundation aims at uniting people who are willing to make a contribution to the growing consciousness of the oneness of all life (www.natuurcollege.nl). They do this to all kind of activities such as retreats and courses, which take part in natural settings.

[9]http://www.forestry.gov.uk/pdf/fce-faith-in-the-woodlands.pdf/$FILE/fce-faith-in-the-woodlands.pdf

Chapter 9
Lessons Learned, Trends and Strategies for the Future

Ulrike Pröbstl

9.1 Lessons Learned

The focus on recreation and landscape aesthetics in forestry is nothing new. In the middle of the nineteenth century the challenge of combining economic and social aspects led to controversy among foresters in central Europe. The earliest discourse focused on forest aesthetics (Hirschfeld 1785, Borch 1824, Salisch 1885, Dimitz 1909), but proponents were not really successful in their call for integrating aesthetic concerns into forest management.

The first concept of European forest management, which prevailed for most of the first half of the twentieth century has come to be known as the "Wake Theory" (Rupf 1960). It is based on the simple assumption that "proper" classical forest management, without any particular concern for any of the other forest functions (e.g. nature conservation, soil conservation, recreation) will automatically result in proper forest conditions and accommodate all the other concerns including recreation.

At the beginning of the 1950s, a new emphasis on research on recreation and other forest functions challenged this approach (see Ammer and Pröbstl 1991, Jensen and Koch 2004, among others).

In the meantime, many European countries have embedded the concept of multi-functionality within their respective judicial and forest policy frameworks, and the majority of countries have implemented relevant functional concepts as documented in the literature and in the survey reported here. In most countries that apply this functional forest classification, the recreational function of the forest must be actively considered in the planning process.

Although the overview presented in this book showed significant differences between the European regions, overall we feel that the different challenges and experiences from each region could also be used to enhance learning from each other

U. Pröbstl (✉)
University of Natural Resources and Applied Life Sciences, Institute of Landscape Development, Recreation and Conservation Planning, Peter-Jordan-Strasse 82, 1190 Vienna, Austria
e-mail: ulrike.proebstl@boku.ac.at

U. Pröbstl et al. (eds.), *Management of Recreation and Nature Based Tourism in European Forests*, DOI 10.1007/978-3-642-03145-8_9,

and to achieve further progress. In the following sections some of these transferable experiences are summarised.

The lessons learned from the Atlantic Region stress its strong focus on a scientifically-based national monitoring of forest recreation. Denmark, the Netherlands as well as the UK are able to base their management decisions on sound social science. One good example in this context is the Great Britain Day Visit Survey (GBDVS) which strongly supports visitor management in the UK.

We also identified a culture of various successful collaborative planning processes, such as the forest user councils established by the Danish Forest and Nature Agency for public involvement in forest management. Private Public Partnerships in Iceland called "Open Forest" fund access to forest areas by the general public, and support the construction of recreational facilities.

Besides monitoring and collaboration, countries in this region are famous for their education and history in incorporating aesthetics into forest planning. More than anywhere else this region provides, guidelines and courses on design principles and outdoor recreation management. The British Forestry Commission, the Irish Tree Council, and the Danish Centre of Forest, Landscape and Planning all offer material which could also be used in other regions in order to be more sensitive in forest management and landscape aesthetics since these concepts have been "exported" to North America more than into other parts of Europe.

The *Atlantic countries* are densely populated and they have – compared to other countries and regions – a high number of immigrants and inhabitants with a minority ethnic background. Therefore one finds in these countries many social empirical studies focusing on ethnic differences in forest recreation and social problems in forests around the urban areas. Good examples, which might be transferable to other metropolitan areas in Europe, e.g. Berlin or Vienna, are the guidelines on vandalism risk management from the UK and recent findings on ethnic differences in forest recreation from The Netherlands.

The lessons learned from the *Nordic region* are closely related to the strong tradition in outdoor recreation management and nature-based tourism in these countries. The tradition of the "friluftsliv", the every man's right and the long tradition of mushroom and berry picking are the basis for new and innovative products and cooperation with tourism. Good examples and best practice in this context are "Nature's Best", a national certification scheme for ecotourism in Sweden, all year round Nordic-Fitness sport concepts or new trails and settings specifically for hiking (e.g. the Skaneleden Trail in Sweden).

In Finland these new and expanded offers are particularly embedded in an attractive unique design targeting nature tourists (Finland). These developments are supported by new research capacity in this field and different forms of cooperation like NOVA – a platform for innovative scientific Nordic co-operation and "Wild North", a new brand of nature based tourism developed by the stated owned forestry company in Finland. Since tourism trends across Europe show a high interest in new nature-based tourism offers, the other regions could profit from these experiences.

The experts in the *Central Region* presented special knowledge in conflict-related research and conflict resolution mostly concerning the natural environment, but also some user conflicts. This experience leads to good examples in this field. In order to stop erosion at the shoreline, to protect the costal environment and the pine forest landscapes more than 30 collaborative zoning concepts for beach plans have been developed in France. Risk management and mapping against natural hazards have been studied and implemented in Germany, France, Switzerland and Austria. For special tasks like managing winter sport areas a Landscape Auditing System for alpine resorts and destinations has been developed, which could serve as examples and good practices for regions and for larger recreational operations. Good practices can be found in Germany, Austria, Switzerland and Italy. Experiences conflict resolution using special systems of signposts in all central European countries. Recently in most of the central European countries, research programmes on the effects of climate change on recreation have emerged. They show the vulnerability of many tourism and recreational offers. Research concepts and their findings could help other countries with their research.

Transferable results of this pan-European survey from the countries of the *Eastern region* are the development of long term planning concepts for forestry and recreation which have been developed in combination with forest inventories (see good practice in Lithuania). Furthermore it is remarkable how old forest infrastructure has been revitalized in order to enhance or establish facilities for recreation. In this context the new interpretation centre and reuse of an old forestry train in the Vydrovo Valley, Slovakia is a good example. New experiences and new products of nature and forest related tourism packages have been developed in several countries, e.g. of eco-tours in Poland.

The strong efforts to enhance forest recreation are also strengthening measures to protect the cultural heritage and landscape. One example for this trend is the ethnographic village in the Dzukija National Park in Lithuania.

If we finally look at the lessons learned in the *Mediterranean region*, we see that they have plenty of experience in the field of tourism. One of the major threats in this context is the risk of fire often caused by tourism activities but also endangering tourism infrastructure. The efforts cover a wide range of educational programmess to increase the awareness of forest fires, e.g. in Croatia and an integrated overall fire management including fire hose reel systems at picnic places in Cyprus and a warning model called Final Risk Index for fire management, modelling the daily risk or seasonal risk for selected areas in Italy. In the past, tourism in many Mediterranean countries has been concentrated on the coastal zone. New transferable concepts which are strongly related to forests will strengthen nature trails and infrastructure in the hinterland and peripheral areas. Cyprus and Portugal especially try to enhance the nature experience for tourists and to attract new target groups.

We also learned that the character and the expectations towards recreational infrastructure in the Mediterranean countries differ significantly from other European regions. Large campsites and areas for picnicking, adapted to local traditions, provide the expected facilities for the local population.

In Italy many of the good practice examples and infrastructure for nature interpretation represent transferable concepts of interest to other European regions.

9.2 Trends and Challenges for the Future

The role of forests for recreation and nature-based tourism is influenced by societal, ecological and economic trends and changes across Europe. In the following we will try to highlight some of them and discuss their possible influence on our subject.

If we just look at the forests, major threats and changes are expected with climate change (Felbermeier 2009). In Central Europe in many regions, in which spruce have been introduced, the precipitation rate will no longer be sufficient to maintain this type of forest in the future. Already global warming is increasing the danger of calamities such as bark beetles especially in monocultural forests. This development is expected to have significant effects on the visual quality of many landscapes and might influence their value for recreation.

The possible effects of global warming on European forests also lead to a discussion on the introduction of foreign tree species like Douglas fir in Central Europe or Eucalyptus in Southern Europe. Portugal has already planted large areas of eucalyptus stands, which dominate the landscape, and negative effects on recreation (especially associated with the smell) have been observed.

New strategies to mitigate climate change, such as afforestation and the cultivation of large fields with energy forest or short-rotation coppice, might also influence the character and the attractiveness of landscapes for tourism and recreation.

Both direct and indirect effects of climate change are also expected to influence recreational behaviour and destination choice. Lack of snow, a loss of glaciers and permafrost regions and an increase of extreme events will lead to negative consequences for tourism and recreation in and around forests. Since forest and land owners are profiting from ski resorts and cross country tracks in many countries this development might increase economic losses. Additional incomes from renting these areas to winter sport enterprises or additional job opportunities in tourism are endangered.

Parallel to these manifold changes around climate change, which are associated with much uncertainty, the forest owners are also confronted with strong legal requirements concerning nature conservation, especially based on the coherent European conservation network "Natura 2000". In most countries, the majority of protected sites are located in forests. The required "avoidance of significant deterioration" by too intensive use or non-use, the consideration of special protected species and the discussion on adequate concepts for contractual or other form of protection, all are typical questions and challenges associated with conservation issues. At the same time, in many countries a discussion on establishing new wilderness areas and recreation-free forests has emerged, all influencing future land use.

Many societal trends will influence the recreational function of forests in the future. In Europe we are facing an ageing society and this trend is already irreversible. Recent studies in tourism and recreation show an increasing demand, driven by elderly people, to enjoy forests and landscape. The increase in accessibility for elderly people and an increasing number of user groups with special needs will be a future challenge for forest management.

By contrast many studies have reported an increasing alienation of young people from nature and forests. So-called "bodyguard parents", who bring their children to school and to other activities by car, leave little uncontrolled time and fewer chances for free nature experiences to be obtained by the young generation, fuel this trend. Furthermore, ethnic influences and the significant role of computer games and other media all accentuate this trend further.

If we look at the trend in spatial development we perceive an ongoing process of urbanisation. The peri-urban area around every European city is expanding. About 70% of the Swiss population already lives in metropolitan areas. This development leads to an increasing demand for different forms of management of the surrounding forests. The provision of infrastructure adapted for different user groups, ranging from dog forests to networks for mountain bikes are required. Recent research has also shown that vandalism and the benefits of forest for socially discriminated people all must be considered in future forest management, suggesting studies on these topics in the future. In many metropolitan areas crowding and the issues of ethnic integration are gaining importance. Compared to the United States, Europe has so far seen far less research on the role of forests for immigrants. The little that has been undertaken in the context of local conflict resolution is related to parks and the built environment. Güleş et al. (2005) studied the preferences of Turkish inhabitants for forests around the industrial zone in central Germany. The immigrants preferred designed, less natural forests compared to the local population. Overall the immigrants keep the tradition of forest use and preferences acquired from their home countries such as cooking in the outdoors. More research is required to increase the understanding about these issues and to enhance the opportunity for conflict resolution and to adapt forest infrastructure to these new demands.

The growth of metropolitan areas and an ever increasing demand for recreation have also had political and legal consequences. Even in parts of the UK new legislation has recently opened more private land for recreation purposes (Dales 2004). This could be a sign for a new challenge for legislation and forestry also in other countries.

Outside the metropolitan areas (i.e. in rural areas) a loss of the younger population is common in most regions. A lack of jobs for those with a higher qualification, deficiencies in public transport and social services are some of the main reasons for an increasing depopulation which can be seen in many peripheral areas of Europe. An enhancement or strengthening of rural tourism is perceived as one possible concept to revitalise these peripheral areas. Therefore many European regions try to develop new nature tourism packages in order to strengthen economically weak peripheral areas.

Finally, in a brief scan of political trends, we perceive an increasing demand for the economic justification and evaluation of recreational benefits, needs and demands. In a political context positive effects of forests for recreation and nature tourism are only taken into consideration if their effect on regional development and local income can be proven. In 2005 for Switzerland the official value of all forest based recreation for all adults has been estimated at about 6 million Euro or €1.095 per person per year (BUWAL 2005).

9.3 Concepts for the Future

The Rio Conference on sustainable development (1992) and the Agenda 21 support the concept of multi-functional forests in the "magic triangle" of environmental, economic and socially balanced circumstances. Currently, sustainable forest management (SFM) has emerged as the overarching concept of forest management in the developed countries worldwide (Gadow et al. 2000), even though different interpretations of the concept exist, and it also overlaps with the concept of ecosystem management (Wiersma et al. forthcoming).

Worldwide frameworks for the implementation of sustainable forest management (SFM) have been developed. A new set of ecological, economic and social indicators and criteria have been defined in order to describe the various functions and benefits of forests in a differentiated manner (e.g. CCFM 2003). These frameworks are typically based on six to nine main criteria. Indicators referring to recreation can typically be found in the context of one very broad criterion such as "maintenance and enhancement of long term multiple socio-economic benefits to meet the needs of societies" or "opportunities for a range of quality-of-life benefits" (see for example USDA 2004, Robinson 2006).

In Europe a sequence of Ministerial Conferences on the protection of Forests in Europe (MCPFE) (e.g. in Lisbon 1998, Vienna 2002, Warsaw 2007) and related expert meetings mainly defined the process and the common understanding of SFM. The target of this pan-European initiative was to widen the understanding of sustainability in forest management, which was originally considered as the sustained yield of timber to cope with historic wood shortages. Furthermore the deterioration of forests throughout central Europe in the 1980s led to an increasing awareness of the economic, social and cultural values of forests. New guidelines also followed the idea that with the implementation of SFM one could translate international commitments down to the level of forest management and practices. From the beginning the framework was also adapted to meet new challenges. Topic such as climate change, biodiversity and socio-economic aspects influenced the discussion and the result of the improvement process. The set of "Improved pan-European Indicators for SFM" was completed in 2003. Since then, work has proceeded on linking SFM to the ECO system approach, developing national forest programmes and researching the role of traditional knowledge in SFM. In fact the MCPFE political commitment involves 45 European countries, European Communities and cooperates with other countries, as well as international organisations that participate as observers. The

European framework is based on 35 quantitative indicators and 12 qualitative ones in six criteria. The criteria are the maintenance and appropriate enhancement of forest resources and their contribution to global carbon cycles (C 1), the maintenance of forest health and vitality (C 2), maintenance and encouragement of productive functions of forests (wood and non wood) (C 3), maintenance, conservation and appropriate enhancement of the biological diversity in forest ecosystems (C 4), maintenance and appropriate enhancement of protective functions in forest management (notably water and soil) (C 5), and maintenance of other socioeconomic functions and conditions (C 6).

Overall the main goals of the concept of sustainable forest management and additional guidelines are:

- to assist forest managers and forest owners in planning and implementing improved sustainable management practices,
- to supervise and control management practices and to support adaptation,
- to bring European commitments down to the field level,
- to enhance a Pan-European and international dialogue,
- to improve communication and awareness related to sustainable forest management,
- to provide references for standards and certification (MCPFE 2003).

The Ministerial Conference on the Protection of Forests in Europe (MCPFE) emphasised that a sound balance of the different indicators is required, which should/must consider the specific local, economic, ecological, social and cultural conditions. But all those who expected that SFM would enhance recreation as one of the main issues were in for a surprise:

Most managers and researchers working in the field of recreation perceive a loss of relevance and miss the above mentioned balance between ecological, economical and social aspects.

Their critique of the criteria and indicator framework centres on two aspects: on the one hand the systematic and structural nature of the criteria and on the other hand the main content. Below, we will elaborate some of these points of critique by focusing on the European concept of SFM. Some of these problems can be found in other frameworks.

Within the long list of indicators (35) only a select few clearly focus on recreation. Therefore one ought to be concerned about the possibility that given the large set of indicators in total, recreation concerns will lose overall importance.

Furthermore, recreational aspects are split into different areas. They can be found under

C 3: Maintenance and Encouragement of Productive Functions of Forest (Wood and Non-Wood) in section

3.3 Non wood goods,
3.4 Services,
3.5 Forests under management plans

and in parts of C 6: Maintenance of other socio-economic functions and conditions in section

6.10 Accessibility for recreation and
6.11 Cultural and spiritual values (number of sites).

The content of the framework also leads to challenges and misunderstanding when applied. Non-wood goods contain mushrooms, cork, berries, nuts, latex, tannins, resins and many other products. Some are actually part of a recreational activity, while others – such as cork clearly are not. The economic value of these non-timber forest products in Europe was 5 billion Euro or 24% of the value of timber in the year 2000. The question is whether the overall recreational function of non-wood goods is actually measurable by economic valuation alone. The combination of various types of values in one good or service may lead to underestimation of the recreational function, especially if the estimation procedure is deficient.

Similar challenges occur if the services must be listed in the monitoring framework. For example, is the guided tour of mountain bikers or the education programme for children in the forest part of the indicator services? Should the income of the guides be a base for calculating the economic benefits, or the number of visitors? If these offers in the forest are free of charge, should these significant recreational offers still be included in the estimation of forest values? In other words, do systematic research frameworks need to be established to account for the consumer surplus accruing from all these recreation activities?

Looking at forest aesthetics and recreation we find in early documents a clear understanding that these aspects must be considered on a practical level. In Annex 2 of the Resolution L2 of the Ministerial Conference on the Protection of Forests in Europe in Lisbon additional information was provided to illustrate the recreation criterion and to provide special guidelines. The guidelines for forest management practices (MCPFE 2003, p. 10) for the criterion "maintenance of other socio-economic functions and condition" (C 6) states that "forest management operations should take into account all socio-economic functions, especially the recreational function and aesthetic values of forests by maintaining, for example, varied forest structures, and by encouraging attractive trees, groves and other features such as colours, flowers and fruits. This should be done, however, in a way and to an extent that does not lead to serious negative effects on forest resources, and forest land".

No criteria are explicitly considering the value of forests for recreation (the influencing factors are well-known and defined by research e.g. species composition, age, structure) are not included. The only focus is on accessibility. The list of improved criteria (MCPFE 2002) underlines that access to forests enables people to benefit from the recreational value of forests which contributes to quality of life. The related explanation points out that many recreational uses are not available or based on legal or effective rights of free access, and therefore this indicator complements any data under indicator 3.3 (non-wood goods) and 3.4 (services) from the social point of view.

Since many countries provide unregulated public access, its relevance must be discussed critically in the future. Consideration of nature-based tourism is another issue. It is expanding in many forests across Europe, a dynamic that has not been foreseen, and is not accommodated accordingly in the frameworks. Given its relevance for regional development in forested areas and local income this issue should receive more attention.

The list of indicators shows the lack of monitoring in many countries. Missing data, like the number of visitors, also lead to the fact that recreation is not represented in that concept in an appropriate manner (Sievänen et al. 2008). Simple statistics, such as visitor numbers provide the basis for calculating their contribution to the local economy (see Job and Metzler 2006). The European court of Auditors, evaluating forestry measures within rural development policy in 2004 has already identified such deficiencies in monitoring. No assessments were made on whether funded measures adequately reflected the three pillars of multifunctionality, economic, social and environmental.

The concept of SFM should be serious about closing this gap in future. Therefore the system should be able to show all aspects in a balanced and comparable manner. In a good monitoring framework, the positive effects as well as the deficiencies of the recreation function should be as obvious as a lack in biodiversity in any one of the SFM frameworks.

Overall it appears that recreation, which was an explicitly defined component in the early implementations of multifunctional forestry, has now lost its special status. It no longer appears as one of the key questions or concerns for management. Therefore politicians and managers implementing these frameworks need to acknowledge the special characteristics of recreation with a unique set of indicators. Even in research the focus has shifted to such concerns as certification, decision support systems, and public participation, which now dominate the discussion in associated research projects. We all hope that this book, especially the Pan-European collection of data about recreational management, will contribute to a better understanding of the role of recreation and an enhancement of SFM. It should be the aim to establish an instrument which is helpful for managers in practice, supervision, information, communication and standardisation.

Members of Working Group «Recreation Planning and Management » within COST Action E33

Country	Name	Office	Address
Austria	Veronika Wirth	University of Natural Resources and Applied Life Sciences, Vienna Institute of Landscape Development, Recreation and Conservation Planning	Peter-Jordan-Str. 82 1190 Wien, Austria
Belgium	Gudrun Van Langenhove	Forest and Nature Agency Flemish Government	Graaf de Ferrarisgebouw, Koning Albert II-laan 20 bus 8 1000 Brussel Belgium
	Pieter Roovers	Aeolus bvba	Vroentestraat 2B B-3290 Diest, Belgium
Croatia	Goran Rubin	Croatian forests Ltd. co. Forest Office Brač	HR-21400 Supetar, Vladimira Nazora 2 Croatia
Cyprus	George Pattichis,	Senior Conservator of Forests Department of Forests, Ministry of Agriculture, Natural Resources and Environment	CY1414 Nicosia, Cyprus
	Savvas Kazafaniotes	Forestry Department	CY1414 Nicosia, Cyprus
Czech Republic	Jan Prochàzka	University of South Bohemia, Faculty of Agriculture,	Studentská 13, 370 05 České Budějovice, Czech Republic
	Jarolav Bohac	University of South Bohemia Faculty of Agriculture	Studentská 13, 370 05 České Budějovice, Czech Republic
Denmark	Berit C.Kaae	Dep. of Urban and Landscape Studies Danish Centre for Forest, Landscape and Planning, KVL	Rolighedsvej 23 1958 Frederiksberg C. Denmark
	Søren Præstholm	Institute of Geography University of Copenhagen	Øster Voldgade 10 DK-1350 Copenhagen K Denmark
Estonia	Toomas Muru	Estonian University of Life Sciences Institute of Agricultural and Environmental Sciences	Kreutzwaldi 5 51014 Tartu, Estonia
Finland	Joel Erkkonen	Metsähallitus, Natural Heritage Services	P.O.Box 8016 (Koskikatu 44-46), 96101 Rovaniemi, Finland

(continued)

Country	Name	Office	Address
France	Peter Breman	ONF Office national des forêts Direction de l'Environment et du Développement Durable	Boulevard de Constance F-77300 FONTAINEBLEAU, France
	Olivier Picard	Institut Develloppement Forestier	23, avenue Bosquet 75007 Paris, France
Germany	Ulrike Pröbstl	AGL-Arbeitsgruppe für Landnutzungsplanung, Institut für ökologische Forschung	St.Andrä-Str. 8, 82398 Etting-Polling, Germany
	Renate Bürger-Arndt	Institut für Forstpolitik, Forstgeschichte und Naturschutz	Büsgenweg 3, 37077 Göttingen, Germany
Greece	Athanassions Hatzistathis	Aristotelian Universtiy of Thessaloniki Dpt of Forestry and Natural Environment Laboratory of Silviculture	54006 Thessaloniki, Greece
Hungary	Dóra Drexler	Corvinus University of Budapest Buda Campus Faculty for Landscape Architecture, Department for Landscape Planning and Regional Development	Budapest 1118 Villányi út 35-43. Building 'K' 2nd floor Room 212, Hungary
Iceland	Jon Geir Petursson	Ministry for the Environment	Skuggasund 1 150 Reykjavík, Iceland
	Sherry Curl	Iceland Forest Service - Forest Planning and Development Div.	Midvangur 2-4 700 Egilsstadir, Iceland
Ireland	William Murphy	Coillte	Newtownmountkennedy, Co. Wicklow, Ireland
	Art McCormack	MosArt Landscape, Architecture, Research	2 Mortons Lane Wicklow County Wicklow, Ireland
Italia	Barbara Mariotti	Università degli Studi di Firenze Dipartimento di Scienze e Tecnologie Ambientali Forestali (DISTAF)	Via San Bonaventura, 13 I-50145 Firenze, Italia
Latvia	Baiba Rotberga	State Forest Service	13. Janvara Iela LV - 1932 Riga Latvija
Lithuania	Olgirda Belova	Lithuanian Forest Research Institute	Liepu 1 Gririonys LT-5310 Kaunas Distr. Lithuania
Netherlands	Jan Blok	National Forest Service	P.O.Box 1300, 3970 BH Driebergen, The Netherlands
	Birgit Elands	Wageningen University & Research Centre Forest and Nature Conservation Policy	Droevendaalsesteeg 3 6708 PB Wageningen, The Netherlands

(continued)

Country	Name	Office	Address
Norway	Jørund Aasetre	Nord-Trøndelagsforskning (North Trondelag Research Institute)	Bomveien 3 N-7725 Steinkjer, Norway
	Odd Inge Vistad	Norwegian Institute for Nature Research	Fakkelgården N- 2624 Lillehammer, Norway
Poland	Agata Cieszewska	Warsaw University of Life Sciences (SGGW), Department of Landscape Architecture	Ul. Nowoursynowska 159 02-786 Warsaw, Poland
Portugal	José Manuel Castro	Agronomic School of Bragança Campus de Santa Apolonia	Apartado 1172 P - 5301-855 Bragança Portugal
	Joao Maria da Rocha Peixoto Cameira	Camara Municipal de Braganca	Forte S. Joao de Deus 5300 Braganca, Portugal
Serbia and Montenegro	Liljana Comic	Faculty of Science University of Kragujevac Serbia	R. Domanovic'a 34000 Kragujevc, Serbia
Slovakia	Maria Bihunova	Department of Garden and Landscape Architecture Faculty of Horticulture and Landscape Engineering Slovak Agricultural University	Tulipanova 7, 949 01 Nitra, Slovakia
	Lubica Feriancova,	Slovak Agriculture University in Nitra, Faculty of Horticulture and Landscape	Tulipanova 7, 949 01 Nitra Slovakia
Sweden	Dan Rydberg	Fastighetskontoret Göteborgs Stad	Box 2258, 403 14 Göteborg Besök: Postgatan 10, Sweden
Switzerland	Marcel Hunziker	Section Landscape and Society Swiss Federal Research Institute WSL	CH-8903 Birmensdorf, Switzerland
	Bruno Baur	Institue of Environmental Sciences Section of Conservation Biology University of Basel	St. Johanns-Vorstadt 10 CH-4056 Basel, Switzerland
United Kingdom	Simon Bell	Open Space Research Centre Edinburgh College of Art	Lauriston Place, Edinburgh, EH3 9DF, UK
	Roger Worthington	Forestry Commission (England)	Bullers Hill Kennford Exeter, EX6 7XR, UK
	James Swabey	Forestry Commission (England)	340 Bristol Business Park Coldharbour Lane Bristol BS16 1EJ, UK

References

Ahlström, I. (2005): Friluftsliv i tätort. Del 2. Underlag för friluftsplan. Exemplet Järvafältet. Länsstyrelsen i Stockholms Län, Rapport 27.

Ammer, U. (1983): Erholung und Landschaft – haben wir des Guten zu viel getan? Forstwiss. Centralblatt, 102(4), 217–233.

Ammer, U. and Pröbstl, U. (1991): Freizeit und Natur, Probleme und Lösungsmöglichkeiten einer ökologisch verträglichen Freizeitnutzung. Pareys Studientexte 72, Hamburg.

Andersson, R. and Rydberg, D. (2005): Naturen och hälsan. Skogsstyrelsens Förlag, Jönköping.

Anon. (2005): Recreation Policy – Healthy Forest, Healthy Nation. Coillte Teoranta – Irish Forestry Board, Ireland. Availabe at www.coillte.ie/publications.

Arnberger, A. and Haider, W. (2005): Social effects on crowding preferences of urban forest visitors. Urban Forestry & Urban Greening, 3(3–4), 125–136.

Arnberger, A. and Haider, W. (2007): Would you displace? It depends! A multivariate visual approach to intended displacement from an urban forest trail. Journal of Leisure Research, 39(2), 345–365.

Bayrische Landesanstalt für Land- und Forstwirtschaft (LWF) (2004): Waldfläche und Besitzverhältnisse, URL: http://2004.lwf.bayern.de/lwfaktuell/lwfakt17/17akt_3.htm, accessed on 07 July 2005.

Bell, S. (1997): Design for Outdoor Recreation. Spon Press, London.

Bell, S. (2007): Design for Outdoor Recreation (Second Edition). Taylor and Francis, London.

Bell, S. Simpson, M., Tyrväinen, L., Sievänen, T. and Pröbstl, U. (2009): European Forest Recreation and Tourism. A handbook. Taylor and Francis, London.

Bell, S., Tyrväinen, L., Sievänen, T., Pröbstl, U. and Simpson, M. (2007): Outdoor Recreation and Nature Tourism: A European Perspective, Living Rev. Landscape Res. 1, URL (cited on 18 January 2008): http://www.livingreviews.org/lrlr-2007-2.

BfN – Bundesamt für Naturschutz (2004): Freiwillige Vereinbarungen Naturschutz Natursport, Ein Leitfaden, BfN-Skripten 106, Bonn.

Breman, P. (2004): Verbal information, COST E33 Edinburgh meeting in November 2004.

BMLFUW – Bundesministerium für Land- und Forstwirtschaft, Umwelt und Wasserwirtschaft (2005): Nachhaltige Waldwirtschaft in Österreich. Österreichischer Waldbericht 2004, Wien.

Borch, W.F. von der (1824): Die Ästhetik im Walde. Sylvan. Jahrb. für Forstmänner, Heidelberg, Leipzig.

Brandt, J. and Vejre, H. (2003): Multifunctional landscapes: motives, concepts and perceptions. In: Brandt, J. and Vejre, H. (eds), Multifunctional Landscapes: Volume I: Theory, Values and History. WIT Press, Southhampton, pp. 3–31.

Buckley, R. (2007): Adventure tourism products: price, duration, size, skill, remoteness. Tourism Management, 28(6), 1428–1433.

Bundesministerium für Verbraucherschutz, Ernährung und Landwirtschaft (2005): Bundeswaldinventur, URL: http://www.bundeswaldinventur.de/enid/4 f.html, accessed on 07 July 2005.

U. Pröbstl et al. (eds.), *Management of Recreation and Nature Based Tourism in European Forests*, DOI 10.1007/978-3-642-03145-8_10,

BUWAL (2004): Waldprogramm Schweiz (WAP-CH) – Handlungsprogramm 2004–2015. Bern, Bundesamt für Umwelt Wald und Landschaft BUWAL.

BUWAL (2005): Der monetäre Erholungswert des Waldes. Econcept im Auftrag des BUWAL. Umweltmaterialien Nr. 193. Bern.

Canadian Council of Forest Ministers (CCFM) (2003): Defining sustainable forest management in Canada – criteria and indicators, http://www.ccfm.org/current/ccitf_e.php (accessed 21.11.2006)

Caspersen, O.H. and Olafsson, A.S. (2006): Oplevelsesværdier og det grønne håndtryk. By og landsplanserien. En metode til kortlægning og udvikling af friluftsoplevelser i Hovedstadsregionen.Center for Skov, Landskab og Planlægning. Landskabsplanserien nr. 27. KVL.

Clark, R. and Stankey, G. (1997): The Recreation Opportunity Spectrum: A Framework for Planning, Management, and Research. U.S. Department of Agriculture Forest Service. Pacific Northwest Forest and Range Experiment Station. General Technical Report. PNW-98 December 1979.

Cole, D.N. and Monz, C.A. (2002): Trampling disturbance of high-elevation vegetation, Wind River Mountains, Wyoming, USA. Arctic, Antarctic, and Alpine Research, 34, 365–376.

Cole, D., Watson, A., Hall, T. and Spildie, D. (1997): High use Destinations in Wilderness: Social and Bio-Physical Impacts, Visitor Responses and Management Options. USDA Forestry Service Research Paper INT.496.

Dales, B. (2004): "New access rights and responsibilities in Scotland", conference paper. Related online version (cited on 07 May 2007): http://www.openspace.eca.ac.uk/costE33/outputs_scotland.htm.

Deutscher Rat für Landespflege (2004): Der Beitrag der Waldwirtschaft zum Aufbau eines länderübergreifenden Biotopverbundes, in: Schriftenreihe des Deutschen Rats für Landespflege, Heft 76, S. 5–28.

Dimitz, L. von (1909): Entwicklung und praktische Ziele der Forstästhetik. Österr. Vjschr. Forstwes., 59(NF27), 115–144.

Directorate for Nature Management (2003): '*Marka*' – Planlegging av by- og tettstedsnære naturområder (*Planning in urban and community forests and nature areas*). Handbook 24. Trondheim, Norway

Directorate for Nature Management (2006): Naturvennlig tilrettelegging for friluftsliv Handbook 27. Trondhiem, Norway.

Elands, B. en Lengkeek, J. (2000): Typical tourists – research into the theoretical and methodological foundations of a typology of tourism and recreation experiences. Mansholt Studies 21, Wageningen University, Wageningen.

Elands, B. en van Koppen, K. (2007): Indicators for the 'Convention on Biodiversity 2010'; public awareness and participation. WOt-werkdocument 53.12, Wageningen.

Felbermeier, B. (2009): Waldbauliche Handlungsmöglichkeiten angesichts des Klimawandels. Vortrag auf der Taugung, Naturpark und Wald, Neuberg an der Mürz, 07 May 2009.

Ferenc, G. (1996): State Forest Management in the Transition Period from Planned Economy to Market Directed One IUFRO-95, Papers and Abstracts.

Fitzpatrick Associates Economic Consultants (2005): Economic Value of Trails and Forest Recreation in the Republic of Ireland. Jointly commissioned by the Irish Sports Council and Coillte. Available at www.coillte.ie/publications.

Forest Service (2006): Forest Recreation in Ireland – A Guide for Owners and Managers. Forest Service, Department of Agriculture and Food, Johnston Castle Estate, Co. Wexford, Ireland.

Forestry Commission (2004): Max Hislop, Mark Twery and Heini Vihemäki; Involving People in Forestry. A toolbox for public Involvement in Forest and woodland planning. Forestry Commission, Edinburgh.

Forestry Commission (2007): Public Opinion on Forestry. Summary results of the 2007 UK Survey of Public Opinion of Forestry, carried out on behalf of the Forestry Commission, Forestry Commission.

Fronczak, K. (2003): Poland's green treasure house, State Forest Information Centre, Bogucki Wyd. Naukowe, Warsaw

Gadow, K. von, Pukkala, T. and Tome, M. (2000): Sustainable Forest Management. Kluwer Adacemic Publishers, Dortrecht.

Güleş, O., Hohn, U. and Keil, A. (2005): Abschlussbericht des Projektes „Stadtnatur – Wahrnehmung, Bewertung und Aneignung durch türkische MigrantInnen im nördlichen Ruhrgebiet unter besonderer Berücksichtigung von Industriewaldflächen. Bochum/Dortmund.

Hägerstrand, T. (1995). A look at the political geography of environmental management. In: Landscape and Life; Appropriate scales for sustainable development. LLAS Working Paper no. 17, University College Dublin.

Hirschfeld, C.C.L. (1785): Theorie der Gartenkunst. Nachdruck der Ausgabe 1782–1785. Hildesheim.

Hoogstra, M.A. and Van Blitterswijk, H. (2002): Financial forest policy instruments in the Netherlands; two examples. In Ottitsch, A., Tikkanen, I. and Riera, P. (eds), Financial instruments of forest policy. European Forest Institute, Joensuu, Finland, EFI Proceedings 42, pp. 117–124.

Hunziker, M., (2000): Einstellungen der Bevölkerung zu möglichen Landschaftsentwicklungen in den Alpen, Birmensdorf, Eidg. Forschungsanstalt WSL. 157 S.

Ingold, P. (2005): Freizeitaktivitäten im Lebensraum der Alpentiere – Konfliktbereiche zwischen Mensch und Tier. Haupt-Verlag, Bern.

Jacob, G.R. and Schreyer, R. (1980): Conflict in outdoor recreation: a theoretical perspective. Journal of Leisure Research, 12(4): 368–380.

Jensen, F.S. and Koch, N.E. (2004): Twenty-five years of forest recreation research in Denmark and its influence on forest policy. Scandinavian Journal of Forest Research, 19(4), 93–102.

Jiricka, A. and Pröbstl, U. (2009): One common way – the strategic and methodological influence on environmental planning across Europe. Environmental Impact Assessment Review (in print).

Job, H. and Metzler, D. (2006): Naturparke + Tourismus = Regionalentwicklung? Natur und Landschaft, 81(7), 355–361.

Kairos future (2005): Kund- och intressentanalys. Slutrapport, Stockholm.

Kajala, L., Almik, A., Dahl, R., Dikšaitė, L, Erkkonen, J., Fredman, P., Jensen, F. Søndergaard, Karoles, K., Sievänen, T., Skov-Petersen, H., Vistad, O.I. and Wallsten, P. (2007): Visitor monitoring in nature areas – a manual based on experiences from the Nordic and Baltic countries. TemaNord 2007:534.

Lidestav, G. and Nordfjell, T. (2003): Med skogsägaren i fokus. LRF Skogsägarna, Stockholm.

Lindenau, G., (2002): Entwicklung der Agrarlandschaften in Südbayern und Ihre Beurteilung durch die Bevölkerung, Franziska Land Verlag, 304 S.

Lindhagen, A. and Ahlström, I. (2005): Visitor studies in nature areas – a manual. Skogsstyrelsen, Jönköping, Rapport 4.

Lundell, Y. (2005): Acces to the forests for disabled people. Skogsstyrelsen, Jönköping, Rapport 1.

Mann, C. (2006): Konflikte in Erholungsgebieten – Ursachen, Wirkungen und Lösungsansätze. Freiburger Schriften zur Forst und Umweltpolitik. Verlag Dr. Kessel, Remagen.

Manning, R.E. (1999): Studies in Outdoor Recreation. Search and Research for Satisfaction (2nd edition). Oregon State University Press, Corvallis, Oregon.

Manning, R.E. (2007): Parks and Carrying Capacity. Commons Without Tragedy. Island Press, Washington.

Marghescu, T. and Anderson, J., (1997): Issues and Opportunities in the Evolution of Private Forestry and Forestry Extension in Several Countries with Economies in Transition in Central and Eastern Europe, FAO, Rome

McCammon, I. and Hägeli, P. (2007): An evaluation of rule-based decision tools for travel in avalanche terrain. Cold Regions Science and Technology, 47, 193–206.

MCPFE – Ministerial Conference on the Protection of Forests in Europe (2002): Improved Pan-European Indicators for Sustainable Forest Management, as adopted by the MCPFE Expert Level Meeting 7–8 October 2002, Vienna, Austria.

MCPFE – Ministerial Conference on the Protection of Forests in Europe (2003): State of Europe's Forest 2003. The MCPFE Report of Sustainable Forest Management in Europe source http://www.unece.org/trade/timber/docs/sfm/europe-2003.pdf

Memento Afocel (2004): Les chiffres cles de la foret privee 2005, accessed 05 May 2005, URL: www.arbocentre.asso.fr/infoFil/images/chiffres-cles2005.pdf

Muhar, A. (2004): Quantification of Public Access to the Landscape: Methods and Case Studies. In: Edinburgh College of Art: Open Space – People Space: International Conference for Inclusive Environments, 27–29 October 2004, Edinburgh.

Pine, B.J. and James, H.G. (1999): The Experience Economy: Work is Theater & Every Business a Stage. Harvard Business Press, Boston, MA.

Pinterits, S. (2003): Die Zugänglichkeit der Landschaft – ein analytischer Vergleich zwischen Österreich und Italien. Diplomarbeit. Universität für Bodenkultur, Wien.

Pröbstl, U., Sterl, P., Visschedijk, P. (2006): Agent-based modelling as a tool to protect endangered habitats and species – Experiences of two model projects in Europe. Presentation at the third International Conference on Monitoring and Management of Visitor Flows in Recreational and Protected Areas, 14–19 October, Rapperswil, Switzerland.

Ricoeur, P. (1994): Imagination in discourse and in action. In: G. Robinson and Rundell, J. Rethinking Imagination: Culture and Creativity. Routlegde, London/New York, pp. 118–135.

Robinson, N, (2006): Developing criteria and indicators of sustainable forest management in the Arrow Forest District. BC. Journal of Ecosystems and Management, 7(1), 50–58.

Rosengren, M. and Franzén, K. (2006): Best practice. Råd och tips vid byggande av besöksanläggningar i naturen. Baltic Sea Region Interreg III B Report.

Rupf, H. (1960): Wald und Menschen im Geschehen der Gegenwart. AFZ, 38, p. 545.

Rydberg, D. and Aronsson, M. (2004): Vår tätortsnära natur. Skogsstyrelsens Förlag, Jönköping.

Salak, B. (2008): Grundlage für ein nachhaltiges Schutzgebietsmanagement im Rahmen der EMAS II-Verordnung, illustriert am Beispiel des Nationalparks Gesäuse, diploma thesis, University of Natural Resources and Applied Life Sciences, Vienna.

Salisch, H. von (1885): Forstästhetik, 3. Auflage 1911. Springer, Berlin.

Schemel, H.J. and Erbgut, W. (2000): Handbuch Sport und Umwelt, 3. Auflage, Meyer und Meyer, Aachen.

Shelby and Heberlein (1986): Carrying Capacity in Recreation Settings. Oregon State University Press, Corvallis, OR.

Sievänen, T. (ed) (2001a): Luonnon virkistyskäyttö 2000 (Outdoor recreation 2000), vol. 802 of Metsäntutkimuslaitoksen tiedonantoja, Helsinki (METLA).

Sievänen, T. (ed) (2001b): Luonnon virkistyskäyttö 2000. Summary: Outdoor recreation 2000. Metsäntutkimuksen tiedonantoja 802, 204 pp. Vammala.

Sievänen, T., Arnberger, A., Dehez, L., Grant, N., Jensen, F.S. and Skov-Petersen, H. (eds) (2008): Forest Recreation Monitoring – A European Perspective. 79, 245; Working Papers of the Finnish Forest Research Institute, Helsinki; ISBN: 978-951-40-2098-8

Statistics Sweden (2004): Fritid 1976–2002. Levnadsförhållanden. Statistiska centralbyrån. Rapport 103.

Supuka, J. and Vreštiak, P. (1984). Zásady tvorby parkových lesov. Bratislava, VEDA.

Tracz, H. and Gaca, J.,(2000): Wpływ presji ruchu turystyczno – rekreacyjnego na szkody w lasach Nadleśnictwa Józefów – Roztocze [in:] Problemy turystyki i rekreacji w lasach Polski, Krajowa Konferencja Naukowa, Warszawa.

USDA Forest Service (2004): National report on Sustainable Forests 2003, FS-766, http://www.fs.fed.us/research/sustain/documents/SustainableForests.pdf (accessed 29.11.2006)

Van Herzele, A., Collins, K. and Heyens, V. (2005): Interacting with green space. Public participation with professionals in the planning and management of parks and woodlands, Flemish Government, Nature and Forest Agency, Brussel.

Vaske, J.J. and Shelby, L.B. (2008): Crowding as a Descriptive Indicator and an Evaluative Standard: Results from 30 Years of Research, Leisure Sciences, 30, 111–126.
Vistad, O.I. (2003): Experience and management of recreational impact on the ground – a study among visitors and managers. Journal for Nature Conservation, 11, 363–369.
Vistad, O.I. (1995): I skogen og i skolten - Ein analyse av friluftsliv, miljøoppleving, påverknad og forvaltning i Femundsmarka, med jamføringar til Rogen og Långfjället. (In the forest and in the head. An analysis of outdoor recreation, recreational experiences, recreational impacts and management in Femundsmarka, compared to Rogen and Långfjället – PhD Thesis in Human Geography – In Norwegian). - Dr.gradsavhandling i geografi, Universitetet i Trondheim.
Vittersø, J., Chipeniuk, R., Skår, M. and Vistad, O.I. (2004): Recreational conflict is affective: the case of cross-country skiers and snowmobiles. Leisure Sciences, 26, 227–243.
Watson, A.E. (1995): An analysis of recent progress in recreation conflict perception of future challenges and opportunities. Leisure Sciences, 17, 235–238.
Wiersma, Y., Duinker, P., Haider, W., Hvenegaard, G., and Schmiegelow, F. (forthcoming): The Relationship Between Protected Areas and Sustainable Forest Management: Many Shades of Green. Sustainable Forest Management Network – State of the Knowledge Report, Edmonton, AB, Canada.
Wöbse, H.-H. (2002): Landschaftsästhetik. Ulmer, Stuttgart.
Wöhrstein (1998): Mountainbike und Umwelt; ökologische Auswirkungen und Nutzungskonflikte. Dissertation der Universität des Saarlandes, Saarbrücken. Pirrot Verlag.

Online Sources and Data bases

Ecotourism Sweden: URL: www.ekoturism.org (accessed: June 2006)
Eurostat: URL: http://epp.eurostat.ec.europa.eu/portal/page/portal/eurostat/home
Exmoor National Park: URL: http://www.exmoor-nationalpark.gov.uk/index/visiting/activities/ cycling/ code_of_conduct.htm, accessed: 13 April 2008.
Finnish Forest Association: Facts from Finnish Forest: URL: www.forest.fi (accessed: October, 2005)
Finnish Forest Research Institute: URL: www.metla.fi (accessed: October, 2005)
Metsähallitus: Outdoor recreation in Finland: URL: www.outdoors.fi (accessed: October, 2005)
Österreichischer Walddialog: URL: www.waldialog.at (accessed: 06 June 2008).
Statistics Sweden: URL: www.scb.se (accessed: June 2006)
Sveaskog: URL: www.sveaskog.se (accessed: June 2006)
Swedish Agency for Economic and Regional Growth: Tourism in Sweden: URL: www.tourist.se (accessed: June 2006)
Swedish Environmental Protection Agency: URL: www.naturvardsverket.se (accessed: June 2006)

Legislation

Countryside and Rights of Way Act (CRoW) (2000): Available online, URL: http://www.opsi.gov.uk/ acts/acts2000/ukpga_20000037_en_1(accessed: January 2008).
Federale Overheidsdienst Economie, KMO, Middenstand en Energie: Available online, URL: (http://statbel.fgov.be/figures/dsp2003_nl.asp) (accessed: January 2008).
Flanders, the Forest Act of 1990: Available online, URL: http://212.123.19.141/ALLESNL/ wet/detailframe.vwp?WETID=-1&SID=0 (accessed: January 2008).
H.M. Government (2000): Countryside and Right of Way Act 2000. Office of Public Sector Information.

Irish Occupiers Liability Act (1995): Available online, URL: http://www.irishstatutebook.ie/1995/en/act/pub/0010/index.html (accessed: January 2008).
Land Reform (Scotland) Act (2003): Available online, URL: http://www.opsi.gov.uk/legislation/scotland/acts2003/asp_20030002_en_1 (accessed: January 2008).
Regulation (EC) No 761/2001 of the European parliament and of the council of 19 March 2001 allowing voluntary participation by organisations in a Community eco-management and audit scheme (EMAS).
Scottish Executive (2003): Land Reform (Scotland) Act 2003. Office of Public Sector Information.
Scottish Outdoor Access Code (2004): available online: URL: http://www.outdooraccess-scotland.com/default.asp?nPageID=324 (accessed: January 2008)

Color Plates

Plate 1 Recreation planning in context with conservation planning
Sveaskog's Ecoparks , Sweden (Rydberg), see box 1

Plate 2 National household survey on forestry, UK (Forestry Commission), see box 4

Plate 3 Public involvement in the management of a nature area "Curringherveld", the Netherlands (Werkgroep C&HEK, Plaatselijk Belang Kornhorn, above, Maranathaschool, Kornhorn, below), see box 6

Plate 4 Active participation in the development and implementation of a recreational area (Terryland Forest Park, Ireland (Bosbeer), see box 7)

Plate 5 Close collaboration between a municipality and a local forestry society (Kjarnaskogur Forest, Iceland (Gunnarsson), see box 8)

Plate 6 Public private partnerships: "open forest" project (Opinn skógur), Iceland (Petursson), see box 9

Plate 7 Forest user councils, Denmark (Lund), see box 12

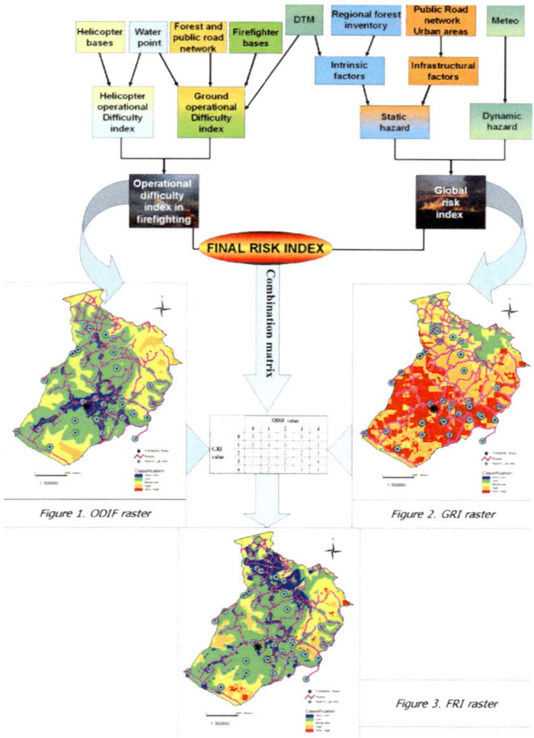

Plate 8 Risk index for fire management in Italy (Regione Toscana – DISTAF University of Florence – CNR IBIMET) (see Box 18)

Plate 9 Risk index for fire management
Fire reel hose system in picnic sites in Cyprus (Kazafaniotis), see box 19

Plate 10 Fire rings in picnic sites, Switzerland and Finland (left, Hegetschweiler) and Finland (right, Paso), see box 20

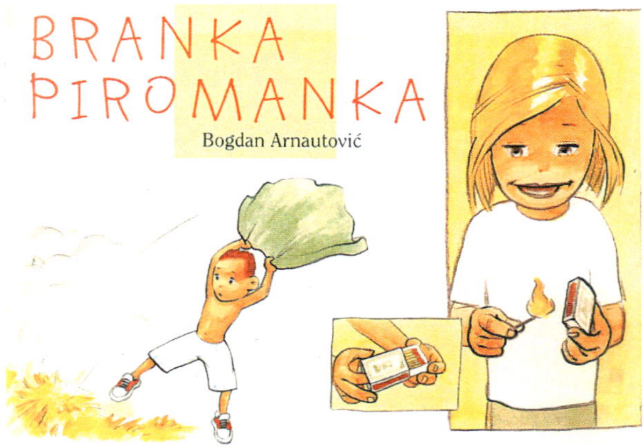

Plate 11 Promoting fire safe behavior to local residents through community based marketing program "I CARE"
Campaign to raise awareness on risk of forest fire (Marusic), see box 21

Plate 12 Fencing off sensitive areas in Troodos National park, Cyprus (Tsintides), see box 25

Plate 13 Wooden trail in an urban forest in Belgium (ANB), see box 27

Plate 14 Wooden trails in Finland
Boardwalk in Evo Hiking Area in Finland (Wirth), see box 28

Plate 15 Voluntary agreements – climbers in Germany (Taczanowska), see box 30

Plate 16 Raising awareness for wildlife disturbance by special user groups
Snowshoe project in the Ibergeregg, Switzerland (Freuler), see box 32

Plate 17 Ranger system to raise awareness and ensure existing rules are kept
Ranger system in Gesäuse Nationalpark in Austria (Thaller), see box 34

Plate 18 Integrated sustainable wildlife management in the urban forest of Vienna (Heckl), see box 36

Plate 19 Separate trails for different user groups
In Jagersborg Dyrehaven Denmark (Kaae), see box 37

Plate 20 Dedication of forests to special uses – dog forests in Denmark (Schiøtz), see box 38

Plate 21 Integrated management: hiking – biking conflicts: The project "Hiking and Biking at the Uetliberg", Switzerland (Freuler), see box 39

Plate 22 Arrangements with specific activity groups
Arrangements with Nordic Walkers in the Netherlands (Kaae), see box 40

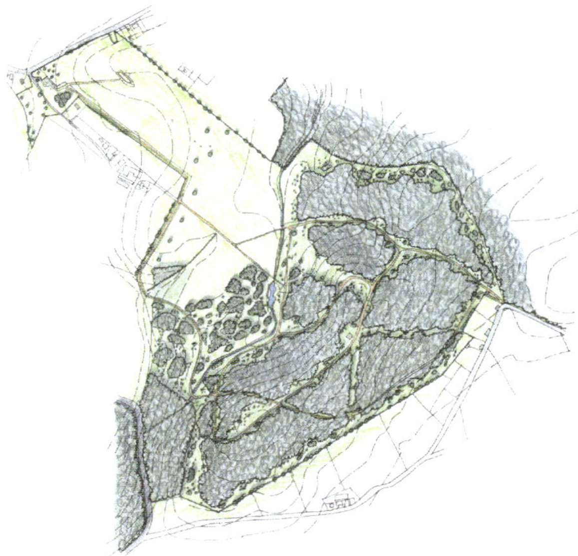

Plate 23 Recreational forest: Victory Wood in England (Bell), see box 43

Plate 24 Recreational Forest of Avondale Forest Park, County Wicklow, Ireland (Bell), see box 44

Plate 25 Aesthetic value: filtered view through forest towards light (Ireland) (McCormack), see box 47

Plate 26 Dramatising the sense of passage at road crest between deep cut banks, see box 52

Plate 27 Mysterious and mystical power of a waterfall in the forest, in the Devil's Glen Wood, County Wicklow, Ireland (McCormack), see box 55

Plate 28 Art in the forest in the Devil's Glen Wood, County Wicklow, Ireland (McCormack), see box 56

Plate 29 Nature's "art" in the forest based on human involvement, Ireland (McCormack), see box 57

Plate 30 Entrance Point to Pyhä-Luosto National Park, Finland (Kuure), see box 59

Plate 31 Entrance point based on traditional Livonian fortress construction (Bell), see box 60

Plate 32 Car park Glen Affric in Scotland (Bell), see box 61

Plate 33 Car park near the coast at Cap Ferret in Aquitaine, France (Bell), see box 62

Plate 34 Information sign at Rosehall community forest, Scotland (Bell), see box 63

Plate 35 Large picnic Site in the Troodos mountains of Cyprus (left: Bell, right: Kazafaniotis), see box 65

Plate 36 Toilets facilities at Abriachan community forest, Scotland (Bell), see box 66

Plate 37 Toilets facilities at in Pykni Forest in Cyprus (Kazafaniotis), see box 67

Plate 38 Provision of play structures in the UK
Large wooden play structures in the UK (Bell), see box 68

Plate 39 Play forests in the Netherlands
Informal play in the Play Forest Mastbos, the Netherlands (Staatsbosbeheer), see box 69

Plate 40 Recreational trail along an abandoned train line, Sabor Ecopista, Portugal (Castro), see box 71

Plate 41 Trails in wetland areas
Boardwalk at Kemeri National Park in Latvia (Bell), see box 72

Plate 42 Trail for disabled use
Trail for wheelchairs at Dyrehaven Forest in Denmark, (Kaae); Trail for wheelchairs at Abriachan community forest, Scotland (Bell), see box 73

Plate 43 Forest access for disabled use in Lithuania (Reseiniai Forest Enterprise), see box 74

Plate 44 Interpretative design for trails
The trail for all senses, Campigna di Santa Sofia, Italy (Giovanni), see box 75

Plate 45 Visitor centre stimulating curiosity and imagination
Visitor centres in the National Park "Foreste Casentinesi, Monte Falterona e Campigna", Italy, (www.parks.it/parco.nazionale.for.casentinesi/), see box 77

Plate 46 Re-using existing facilities as visitor centre
Visitor centre of the Athalassa National Forest Park, Cyprus (Kazafaniotis), see box 78

Plate 47 Visitor centre with nice architecture
Visitor centre at Glendalough (Ireland) (Bell), see box 79

Plate 48 Mobile steel visitor centre, New Forest, UK (Worthington) see box 81

Plate 49 Picnic, shelter and fire place in the Arctic Circle Hiking Area in Finland (Paso), see box 82

Plate 50 Shelter in Rosehall community forest, Scotland (Bell), see box 83

Plate 51 Picnic Shelter in Evo Hiking Area, Finland (Bell), see box 84

Plate 52 Historic Shelter Budakeszi Wildlife Park (Anon) see box 85

Plate 53 Adventure Park of Val di Vara, Italy (Mariotti), see box 87

Plate 54 Open Air Museum of Gyftocampos, Greece (Aristotelian University of Thessaloniki), see box 88

Plate 55 Fairy Tale forest and Dwarf wood, Tervete Country Park, Latvia (Latvijas valsts meži), see box 89

Plate 56 Live role play in Denmark (Andersen), see box 90

Plate 57 Bare feet path (Buro Kloeg, the Netherlands), see box 91

Plate 58 Nordic Walking trails in Finland (Erkkonen), see box 92

Plate 59 Trim trail at Athalassa National Forest Park, Cyprus (Kazafaniotis), see box 93

Plate 60 Therapeutic use of the forest (Kaae), see box 94

Plate 61 Smell and feeling boxes with info boards in serigraphy and Braille (Raes), see box 95

Plate 62 Teaching school children about forests
Forest School, Hungary, (Erdeiiskola), see box 97

Plate 63 Outdoor in all weather nurseries (Rydberg), see box 98